Missions from JPL
Fifty Years of Amazing Flight Projects

Robert Aster
October, 2009

Dedication

More than a thousand dedicated people are needed to build and conduct a mission into outer space. The larger projects at JPL have compiled lists of over four thousand individuals. Many tens of thousands of people have worked hard to make all of these missions succeed. This book is dedicated to all of these professionals.

The author has tried to identify some of the more significant leaders of each mission, and all of the major contributing organizations. A great many other organizations have contributed instruments, spacecraft, launch vehicles and operational services to the JPL missions. This list of contributors is most likely incomplete in some areas, but it is the best compilation of contributing organizations and key leaders that the author can provide.

Some of the images in this book are historical. Some are not readily available except through scanning or screen copying. Every effort has been made to provide the best quality material, but some compromises were necessary with a few of the images.

I hope you enjoy these amazing flight projects from JPL.

The Author

Introduction

This is a pictorial history of missions launched in the first half century (1958 to 2008) of the Jet Propulsion Laboratory (JPL). JPL is a federally funded research and development center dedicated to the robotic exploration of space. It is managed for NASA by Caltech. It sits in the foothills of the Angeles mountains, in northwest Pasadena (see picture below)

JPL primarily conducts science missions. It has the expertise needed to take missions from initial design to development and throughout operations and data analysis. JPL starts with questions posed by the science community; conceives a mission that provides answers to the questions; matures the technologies that enable the mission; designs and develops the mission; operates the mission; and receives, processes, analyzes, disseminates, and archives the data. Throughout this entire process, JPL works closely with the scientific and industrial communities and actively engages the public. This is the complete end-to-end capability that defines the character of JPL.

"The Jet Propulsion Laboratory is a national treasure with an amazing record of successes unmatched in the world."

> Dr. Ed Weiler
> Associate Administrator
> Office of Space Science, NASA

You will find every flight project that JPL has launched for NASA through 2008 in this book. Below is a picture of JPL. My office is right in the middle of things (the Author).

Credits: Grateful acknowledgement is made for the images that are provided courtesy of NASA/JPL-Caltech, which appear throughout this book.
The cover shows (clockwise): America's first satellite; the Mars Exploration Rover; colliding galaxies; Deep Impact striking a comet in space.

Table of Contents

JPL Begins

The California Institute of Technology (Caltech) started the Guggenheim Aeronautical Laboratory in 1926, under the leadership of Caltech professor Theodore von Kármán.

The initial investment in facilities and equipment was $300,000 with an annual operating budget of $15,000. Von Kármán's team did pioneering work in the extremely challenging field of rocket propulsion. Von Kármán was head of Caltech's Guggenheim Aeronautical Laboratory through 1944. Several of his graduate students and assistants performed rocket experiments at Caltech until enough explosions had occurred to earn them the nickname "the suicide squad", and they were ordered off-campus.

They gathered to test a primitive rocket engine in a dry riverbed wilderness area in the Arroyo Seco, a canyon north of the Rose Bowl in Pasadena, California. Their first rocket firing took place on October 31, 1936. In the photo to the right, the participants are tentatively identified as Jack Parsons (wearing the dark vest), Frank Malina (at right), and Rudolph Schott (in short sleeves). This photo appears to have been taken during one of the November 1936 tests. By January 16, 1937 they had achieved a record of 44 seconds of successful rocket engine firing.

Funding by Army Ordnance started in 1938. At this time the Laboratory was a team of about 100 engineers. An early breakthrough was the development of a more reliable solid fuel rocket. This led to the development of "jet-assisted take-off" (JATO) rockets to help overloaded Army airplanes taking off from short runways. Manufacturing JATOs was then spun off to the newly formed Aerojet Corporation. The famous Doolittle raid on Japan was made possible by JATOs. The picture to the right is an experiment conducted just prior to the Doolittle raid.

The name "Jet Propulsion Laboratory" (JPL) first appeared on a report from professor von Kármán on November 20, 1943.

JPL expanded and the team began testing 8 foot tall, unguided missiles (named the Private) that reached a range of nearly 11 miles. The first successful use of booster rockets occurred at this time. **Starting in August 1944, JPL built the Private A.**

The picture to the left shows the Private A test site near Barstow, California. The Private A was a two-stage solid fuel missile. In this photograph the main stage is being hoisted into the rail launcher in December 1944.

They experimented with radio telemetry from missiles, and began planning for ground radar and radio sets.

By 1945, with a staff approaching 300, the group had begun to launch a new set of test vehicles from White Sands, New Mexico, to an altitude of 200,000 feet, monitoring performance by radio. Control of the guided missile was the next step, requiring two-way radio as well as radar and a primitive computer (using radio tubes) at the ground station. The result was named Corporal, a 39 foot tall missile, which first launched in May 1947.

The picture to the right shows a Corporal and a Private missile. The first full sized Corporal was launched in May, 1947.

Technology evolved substantially over the next few years as JPL mastered the fine points of rocket technology, including guidance and control. The first Corporals could carry a payload of 300 pounds approximately 62 miles. This evolved to a 1000 pound payload, with a range of approximately 200 miles.

In 1951 JPL selected Firestone Tire and Rubber Company to build Corporal missiles for JPL. When completed, these missiles were delivered to the JPL Missile Assembly Laboratory, a building which still exists today and is seen in the 1955 photo below.

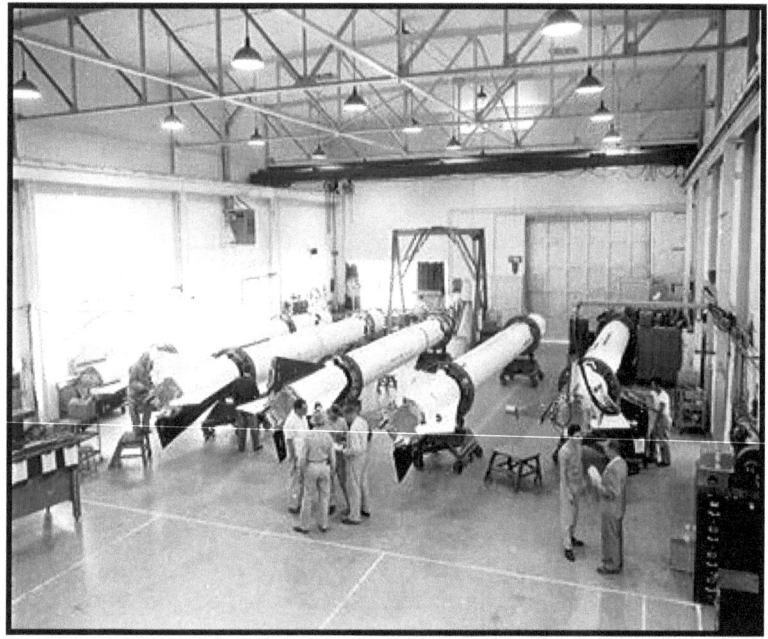

Developing these complex devices to fly unaided and beyond reach of repair meant a new degree of quality, new test techniques and a new discipline called system engineering. Subsequent Army work further sharpened JPL's knowledge of communications and control technology.

The final and most advanced system developed by JPL for the Army was the Sergeant missile.

These technological advances made it possible for JPL to develop the flight and ground systems and finally to fly the first successful U.S. space mission, Explorer 1.

Explorer – America's First Satellite

Explorer was a response to Russia's Sputnik mission. The Army Ballistic Missile Agency (ABMA) provided a Redstone missile first stage. JPL provided a cluster of 11 Sergeant solid fuel rockets for the second stage, four Sergeants for the third stage, and a single Sergeant for the final stage. This system had already been tested, which enabled JPL and ABMA to assemble and launch Explorer 1 in just 84 days.

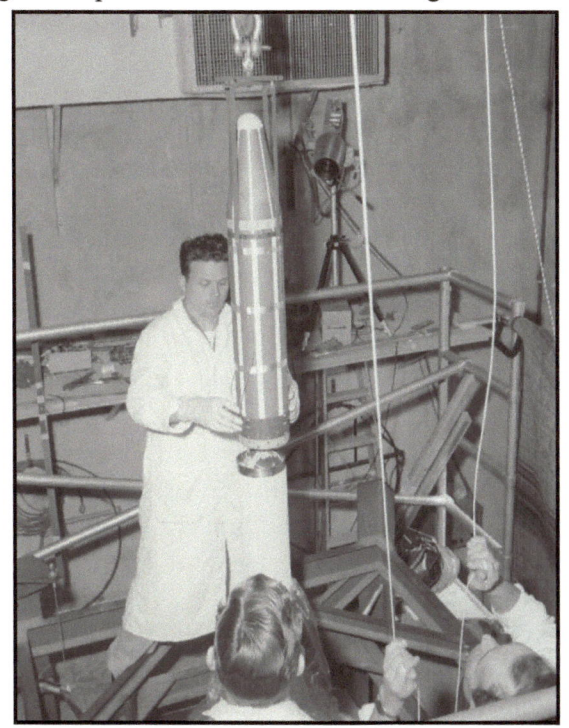

Professor Van Allen of Iowa University provided a scientific instrument that measured charged particles and cosmic rays. The Air Force Cambridge Research Center provided a micrometeorite experiment.

In this photograph, technicians lowered the Explorer 1 satellite onto the launch vehicle's fourth stage motor. The entire spacecraft weighs just 30.8 pounds.

The photo was taken in the gantry at Patrick Air Force Base, Florida, on January 20, 1958.

The Back Story

The story of the decision to launch a U.S. Army earth satellite goes back to 1955. It began as a joint operation of the Jet Propulsion Laboratory, managed by William Pickering, and the Army Ballistic Missile Agency, managed by Wernher von Braun.

There was great interest in launching a U.S. earth satellite as part of the observance of the International Geophysical Year. The Department of Defense appointed a committee to look into existing hardware and at new plans. All three services: the Army, Navy, and Air Force proposed separate satellite launchers. The decision was made to give the job to the Navy Vanguard Project.

After the decision was made in favor of Vanguard, JPL and ABMA (which later became the Marshal Spaceflight Center) redesigned their original Jupiter C proposal into a re-entry test vehicle (RTV). The RTV program goal was to design a nose cone capable of standing the aerodynamic heat as the missile left space and reentered the atmosphere of the earth. In September of 1956, ABMA and JPL successfully completed the first test flight of Jupiter C. The missile reached an altitude of 650 miles and landed more than 3000 miles away. This system was capable of putting a satellite in orbit, but this capability was not utilized due to the assignment of that duty to the Vanguard Project.

Sputnik flew on October 4 of 1957. Soon after that, the Department of Defense asked the Army to present new proposals for an Army satellite. It was proposed that the Jupiter C system be combined with the Van Allen instrument package that was also part of the Vanguard program. This proposal was accepted.

Dr. Pickering selected Dr. J. E. Froehlich to lead the JPL effort. Under the agreement worked out between JPL and ABMA, ABMA was responsible for the operation and launching of the first stage, a modified Redstone liquid-propellant rocket. JPL was responsible for the second, third and fourth stages of the launch vehicle, the satellite, the satellite instrumentation, and collecting information gathered by the satellite.

From the time the Department of Defense directed ABMA and JPL to go ahead and the date of the firing--set jointly by ABMA and JPL--there was only 84 days. But thanks to the early development work on the Jupiter C as an RTV this was enough time. The satellite itself was 80 inches long, and 6.25 inches in diameter, and weighed a total of 30 pounds.

Explorer launched on January 31, 1958 (see figure to the right). Explorer 1 revolved around Earth in a looping orbit that took it as close as 220 miles to Earth and as far as 1,563 miles. It made one orbit every 114.8 minutes, or a total of 12.54 orbits per day. Explorer 1 made its final transmission on May 23, 1958. It entered Earth's atmosphere and burned up on March 31, 1970 after more than 58,000 orbits.

Consequences
Van Allen discovered the Van Allen radiation belts, the first scientific discovery of the space era. The United States had successfully entered the space race with the Soviets.
A jubilant William Pickering, James Van Allen and Wernher von Braun (from left) hoisted a model of Explorer 1 for the cameras at a post-launch news conference.

A Note on "Bill" Pickering

Caltech president DuBridge appointed Pickering director of JPL in August 1954. Pickering stayed in this post 22 years, retiring April 1, 1976. Pickering wanted very much to move the lab out of the weapons business during the late 1950s, and the Soviet launch of Sputnik gave him the opportunity.

Shortly after T. Keith Glennan was appointed head of the newly created National Aeronautics and Space Administration (NASA), he met with Lee DuBridge, President of Caltech, to discuss the transfer of JPL from the Army to NASA.

In the photo to the right are: (L to R) DuBridge; Glennan; Clark Millikan, Director of Caltech's Guggenheim Aeronautical Laboratory; Hugh Dryden, Glennen's Deputy Administrator; and William Pickering, Director of JPL. In December 1958, JPL became the only NASA center managed by an educational institution.

JPL participated in four more Explorer launches (Explorer 2 through 5). Explorer 2 was launched on March 5, 1958, but the fourth stage of the Jupiter-C rocket failed to ignite. Explorer 3 was successfully launched on March 26, 1958, and operated until June 16 of that year. Explorer 4 was launched July 26, 1958, and operated until October 6 of that year. Launch of Explorer 5 on August 24, 1958, failed when the rocket's booster collided with its second stage after separation, causing the firing angle of the upper stage to be incorrect.

Pioneer 3 and 4 – Building on the Explorer Heritage

Pioneer 3 and 4 were early satellites designed to fly by the Moon, with the goal of beating the Soviet Union to become the first Moon mission. They were equipped with Geiger counters to measure radiation in space, much like the Explorer.

Pioneer 3 was launched from Cape Canaveral, Florida, on December 6, 1958. Because of a slight error in the satellite's velocity and angle after burnout of the Juno II rocket, it did not reach the Moon; instead it achieved a peak altitude of 63,580 miles. The satellite discovered a second radiation belt around Earth during its flight, and it validated the new communications technology at JPL that became the Deep Space Network. Pioneer 3 reentered Earth's atmosphere over equatorial Africa a day after launch.

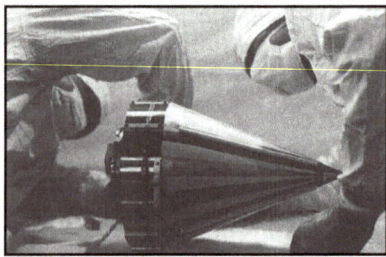

Pioneer 4 was launched March 3, 1959, and successfully passed within 37,300 miles of the Moon the following day (twice the planned flyby altitude). The satellite was tracked for 82 hours to a distance of 407,000 miles from Earth, a record at that time. Pioneer 4 is now orbiting the Sun, the first U.S. spacecraft placed in solar orbit.

Consequences

Pioneer 3 improved our knowledge of the Van Allen Belts, and Pioneer 4 provided data on the Moon's radiation environment.

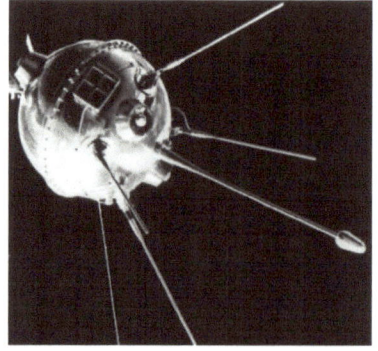

However, the desire to be the first man-made vehicle to fly past the moon was lost when the Soviet Union's Luna 1 (shown to the right) launched on January 2 and passed by the Moon several weeks before Pioneer 4.

It was clear that the Soviet program was still ahead of the American space program in 1959.

Transition to NASA

The National Aeronautics and Space Administration (NASA) started in 1958, and JPL was transferred from the Army to NASA on January 1, 1959.

JPL began development of the Deep Space Network in 1958, in anticipation of the Lunar and planetary missions that were on the horizon. The picture below shows the first 26-meter steerable parabolic dish antenna, in the Mojave Desert, which was begun in the summer of 1958 and completed in November of the same year.

JPL completed development work on the Sergeant missile in 1960, transferring responsibility for the deployment phase of this program to an aerospace contractor (Sperry) which reported directly to the Army. This effectively ended the launch vehicle work of the Laboratory, after three decades of spinning off breakthroughs in solid and liquid fueled rockets.

But the big challenge facing JPL was a transition from a technology development laboratory that could build many prototypes, in search of break-through technologies, to a laboratory that could build a 'first of its kind' spacecraft that worked perfectly the first time it launched. This was both a technical and a management challenge.

13

Rangers 1 through 9 – Failure and Redemption

The Ranger project of the 1960s was the first U.S. effort to launch probes directly toward the Moon. The spacecraft were designed to relay pictures and other data as they approached the Moon and finally crash-landed into its surface.

Ranger projects were grouped into 3 distinct blocks. Blocks 1 and 2 were managed by Cliff Cummings. Block 1 had Rangers 1 and 2, which emphasized the study of the radiation environment in space.

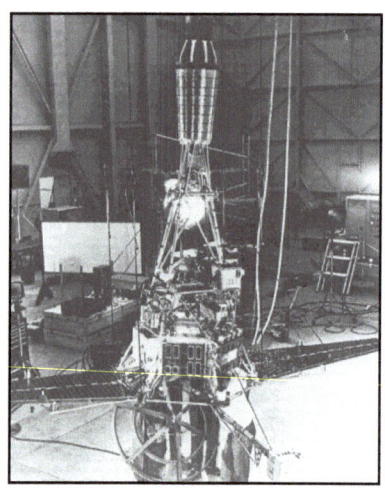

Ranger 1 was launched from Cape Canaveral, Florida, on August 23, 1961. This was followed by the launch of Ranger 2 on November 18 of that year. In both cases, the mission failed when the Agena rocket engine failed to restart and both spacecraft reentered Earth's atmosphere a short time later. The picture to the right shows Ranger 1 during final assembly at the Cape.

Block 2 had Rangers 3, 4 and 5. Each of these missions failed.

Ranger 3 was launched January 26, 1962, but it missed the Moon. Ranger 4 launched perfectly on April 23, but the spacecraft was completely disabled. The picture to the right is Ranger 4, in the assembly hangar at Cape Canaveral. Technicians are preparing the spacecraft for launch. An impact absorbing sphere made of balsa wood sits atop the spacecraft, painted with a saw-tooth pattern to maintain thermal balance during its mission to the Moon. The sphere contained a lunar seismometer, which was to rough land just south of the equator on the rim of the Ocean of Storms and measure "lunar-quakes."

The master clock in Ranger 4's computer failed during flight and the spacecraft did not respond to commands. It crashed into the far side of the Moon on April 26, 1962. The project team tracked the seismometer capsule to impact just out of sight on the far side of the Moon, validating the spacecraft's communications and navigation system.

Ranger 5 missed the Moon following its launch on October 18, 1962, and was disabled. At this point the Ranger program paused to regroup.

Block 3 was managed by H. M. "Bud" Schurmeier. Ranger 6 was launched January 30, 1964. It had a flawless flight culminating in impact as planned on the Moon, but its television system was disabled by an in-flight accident and could take no pictures.

The last three Rangers were completely successful. Each spacecraft carried a total of six television cameras, which were a new design by Dr. Kuiper of the University of Arizona, built

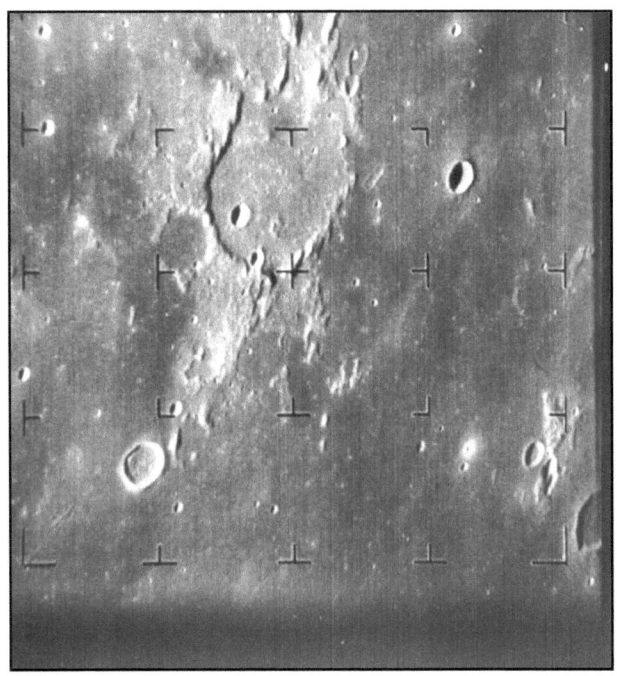

by RCA's Astro-Electronics Division. Ranger 7 was launched July 28, 1964, and hit the surface of the Moon three days later, at the Sea of Clouds.

During the 17 minutes leading up to impact, Ranger 7 transmitted 4,316 photographs to Earth. This image came from Ranger 7.

Following its launch on February 17, 1965, Ranger 8 successfully completed its mission with a planned crash-landing in Mare Tranquillitatis, where the Apollo 11 astronauts would land 4 and a half years later. Ranger 8 returned more than 7,300 images.

Ranger 9 was launched March 21, 1965, and hit the Moon in the 75 mile-diameter crater Alphonsus, sending back more than 5,800 images.

Results

Ranger paved the way for the manned lunar program, providing the best imagery of the lunar surface to date. Very important contributions of Ranger were technology development (including but not limited to improved navigation, improved payload technology, and advances in spacecraft design) and improvement of project management methods.

Mariner 1 and 2 – The First Successful Flyby of Venus

Mariner 2 was the world's first successful interplanetary spacecraft. It was managed by Jack James, who became the prototypical JPL project manager for many years, based on his Mariner performance. R. C. Wyckoff was the project scientist.

The science objectives were: (1) to better understand the only other planet that is similar to Earth in terms of size and (2) to measure the magnetic fields and density of particles between Earth and Venus. The payload came from MIT, Caltech, Brigham Young University, Rice University, Baylor University, the University of Iowa, the University of Nevada, JPL, and the Goddard Spaceflight Center (GSFC).

Two nearly identical spacecraft were launched to Venus in order to improve the odds of mission success. In the early years of space flight, mission failure was more common than mission success. The picture to the right shows the Mariner spacecraft being assembled.

Mariner 1 was launched on July 22, 1962. The rocket carrying Mariner 1 went off-course during launch, and was blown up by a range safety officer about 5 minutes into flight. The picture to the left shows the launch failure in its initial stages. Notice the asymmetrical flame around the bottom the launch vehicle.

Mariner 2 was launched successfully on August 27, 1962, sending it on a three and a half month flight to Venus. On the way it measured for the first time the solar wind, a constant stream of charged particles flowing outward from the Sun. It also measured interplanetary dust, which turned out to be more scarce than predicted. In addition, Mariner 2 detected high-energy charged particles coming from the Sun, including several brief solar flares, as well as cosmic rays from outside the solar system.

On December 14, 1962, Mariner 2 passed within 20,900 miles from the surface of Venus. It scanned the planet with infrared and microwave radiometers, revealing that Venus has cool clouds and an extremely hot surface. (Because the bright, opaque clouds hide the planet's surface, Mariner 2 was not outfitted with a camera.) It discovered that Venus lacks a strong magnetic field and radiation belts, and that Venus' surface temperature was over 400 deg C.

Mariner 2's signal was tracked until January 3, 1963. The spacecraft remains in orbit around the Sun.

This is an artist's concept of Mariner 2 in flight. The round structure to the right is the high gain antenna for communication to Earth across millions of miles of space. The flat panels in the middle are the solar arrays. The box underneath the panels contains most of the electronics of the spacecraft. The structure to the left holds most of the science instruments.

What is a Radiometer?
Physical materials emit small amounts of electromagnetic radiation at certain wavelengths peculiar to each type of material. A radiometer is a device for measuring the intensity and wavelength of this radiant energy with enough accuracy that a scientist can identify certain types of material being observed. A radiometer operates within a certain band of the electromagnetic spectrum. Infrared and microwave are two of the more common bands of the electromagnetic spectrum that are measured by radiometers. Each band has its own advantages for identifying different types of material.

Results
First, JPL engineers learned that interplanetary flight and scientific exploration were feasible using relatively small spacecraft launched by medium-sized launch vehicles. They could be developed in a few years, and they could survive in space for at least a few years.

Second, deep space communication was successfully demonstrated. The Mariner 2 radio transmitter only generated 3.5 watts of power, but JPL's Deep Space Network was able to capture this signal.

Finally, scientists learned a great deal about the atmosphere of Venus and about the electromagnetic fields and particles in the solar system between Earth and Venus. Venus proved to be surprisingly different from Earth. It has a dense carbon dioxide atmosphere topped by marked, opaque clouds and covering a hot surface made even hotter by the atmosphere's "greenhouse effect."

The Mariner program provided a series of projects that pioneered the management processes that led to JPL's success on many subsequent missions. This was the first time a NASA mission accomplished its objective ahead of the Soviets. The Soviets had launched their first Venus mission on February 4, 1961, but it failed to leave Earth orbit.

The Back Story
The management of a flight project is just as important to mission success as the engineering. In the 1940s and 1950s JPL had utilized a program management methodology known at the time as "the arsenal method". This methodology is ideal for aggressive, long-term technology development, featuring the construction and testing of a large number of prototypes that incrementally improve a technology. Each prototype is often tested to the point of failure, yielding more data that quickly leads to a better subsequent prototype. JPL launched more than a hundred Corporal missile systems in order to develop long-range missile technology.

But the cost and the visibility of the NASA missions were too high to use the arsenal method. The Ranger management approach was the first really big departure from the arsenal method, which had provided the technology for Explorer and Pioneer.

But developing a new and successful management approach is not easy. Blocks 1 and 2 of the Ranger program were considered a management failure of sufficient magnitude that the future of JPL in the space program was in jeopardy. The management system that emerged was the project life cycle and the project organization that, in essence, is still used today. Jack James (who

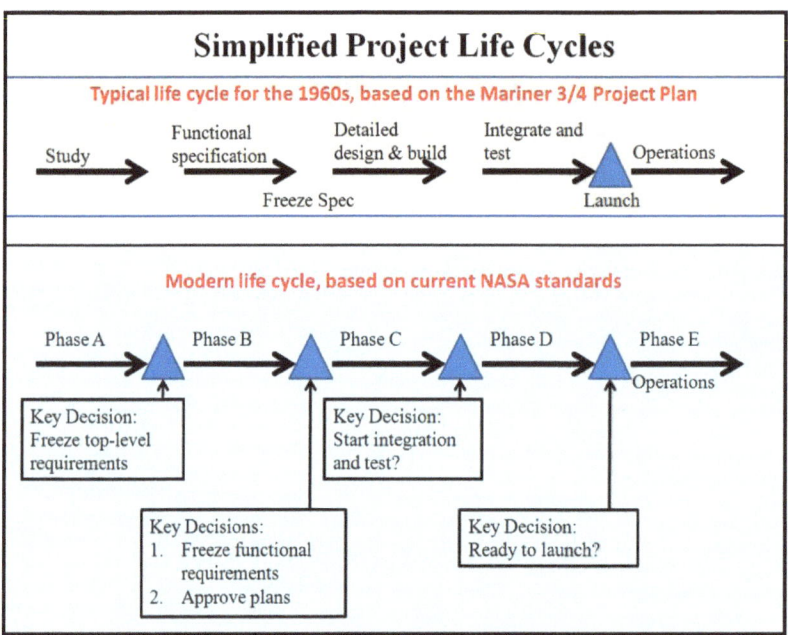

led the early Mariner missions), Bud Schurmeier (Ranger Block 3 manager) and Gene Giberson (Surveyor manager) were the project management pioneers of JPL. The figure to the right shows the early Mariner life cycle and the current NASA life cycle.

Mariner 3 and 4 – NASA's First Successful Flyby of Mars

Mariner 3 and 4 was NASA's first mission to the "red planet". It was managed by Jack James until Mariner 4 launched, and by Dan Schneiderman after that. This began a pattern of management rotation that became typical for JPL missions. Bruce Murray, who would eventually become the Director of JPL, worked on this mission.

The science objectives were to: (1) obtain the first close up pictures of the Martian surface and (2) study the Martian atmosphere. Instruments came from Caltech, JPL, MIT, the University of Chicago, the University of Iowa, and Baylor University.

Two identical spacecraft were launched to fly past Mars, plus a prototype spacecraft. Each

spacecraft carried a television camera, a helium magnetometer to measure magnetic field strength, an ionization chamber and particle detector to measure radiation, a cosmic dust detector, a cosmic ray telescope to measure charged particles, a "trapped radiation detector", and a solar plasma probe to measure the density, velocity, temperature, and direction of movement of low-energy protons from the Sun. The picture to the right shows Mariner 4.

Mariner 3 was launched on November 5, 1964, but the shroud encasing the spacecraft atop its rocket failed to open properly and Mariner 3 did not get to Mars.

Three weeks later, on November 28, 1964, Mariner 4 was launched successfully on an eight-month voyage to the red planet. Mariner 4 passed Mars at a distance of 9,846 kilometers on July 14, 1965, collecting the first close-up photographs of another planet. The pictures, played back from a small tape recorder over a long period, showed lunar-type impact craters, some of them touched with frost in the chill Martian evening.

The 22 photos Mariner 4 sent back from Mars covered about 1 percent of the Martian surface. They show a cratered, ancient world more like the Moon than Earth. To the right is the eleventh Mariner 4 image of Mars. It shows the 151 km diameter Mariner crater (named after the

Mariner spacecraft). The image above was taken from 12,600 km and covers 250 km by 254 km. North is up.

The spacecraft also found that the Martian atmosphere is very thin and unlikely to harbor life. Once past Mars, Mariner 4 orbited the Sun prior to returning to the vicinity of Earth again in 1967. Engineers then decided to use it for a series of operational and telemetry tests to improve their knowledge of the technologies that would be needed for future interplanetary spacecraft. The Mariner 4 spacecraft was expected to survive for a little more than eight months in order to fly by Mars, almost three times longer than Mariner 2. It actually lasted nearly three years.

In July 1965, at a White House ceremony, President Lyndon B. Johnson was presented with a copy of Mariner 4's famous picture number 11, which revealed large impact craters and other topographical features of Mars.

From left are Dr. William H. Pickering, Director of JPL; Oran Nicks, NASA's Director of Lunar and Planetary Programs; Jack N. James, JPL's Mariner 4 Project Manager; President Johnson; and James Webb, NASA Administrator.

Consequences
People used to think that Mars was very Earth-like. Prior to Mariner 4, many people speculated about Martian canals and even Martian civilizations. This mission provided close-up observation of the real planet Mars, which turned out to be very different from the imaginary Mars. This began a process of discovery about planets and planetary climates that will eventually enable scientists to understand our own home in the solar system.

Once again a NASA mission accomplished its objective ahead of the Soviets. The Soviets had launched their first two Mars probes on October 10, 1960, but both failed to leave Earth orbit.

Surveyors 1 to 7 – The First Lunar Landers

The Surveyor series was the first U.S. effort to make a soft landing on the Moon. The purpose of the seven Surveyor missions was to land safely on the Moon, test the landing techniques planned for the manned Apollo landers, take close-up images of the surface, and find locations that would be safe for Apollo landings. Five of the spacecraft successfully soft-landed and returned a great quantity of data, accomplishing all mission and project objectives.

Surveyor was considered at that time to be one of NASA's greatest technical risks, vastly more complicated than Ranger. Instruments came from Caltech, the U.S. Geological Survey, and the University of Chicago. The spacecraft was built by the Hughes Aircraft Company.

Surveyor 1 was launched from Cape Canaveral, Florida, on May 30, 1966, settling down on the Moon's Ocean of Storms - just 46 feet from its intended target on June 2. Surveyor 1 sent 11,240 pictures, revealing details as small as 1/12th inch. The lander operated until January 7, 1967. The picture to the left is the shadow of Surveyor 1, on the Moon.

Surveyor 2 launched on September 20, 1966, but crashed into the Moon three days later.

Surveyors 3, 5, 6 and 7 repeated the triumph of Surveyor 1 in different sites and successively added a robot arm with scoop and a chemical element analyzer to the scientific toolkit.

Surveyor 3 was launched April 17, 1967, and operated on the Moon until May 4, 1967. The Surveyor 3 landing site became the destination for Apollo 12, which arrived in November 1969.

The picture to the right shows Surveyor 3 on the moon, analyzing a sample of lunar soil.

The picture to the right was taken by an astronaut two and a half years later. The Apollo 12 lunar lander can be seen in the background.

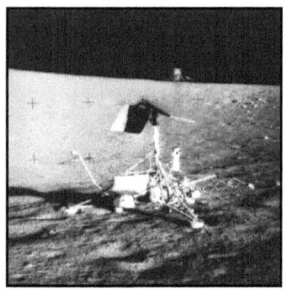

Surveyor 4 was launched July 14, 1967, but its signal was lost 2-1/2 minutes after lunar impact.

Surveyor 5 was launched September 8, 1967 and lasted until December 17.

Surveyor 6 was launched November 7, 1967 and operated until December 14, 1967. Its landing site was in Sinus Medii, which is in the center of the Moon's visible hemisphere. This is the last of four potential Apollo landing areas designated for investigation by the Surveyor program.

Surveyor 7 was launched January 7, 1968 and lasted until February 21 of that year.

The picture on the right shows a lunar panorama near the Tycho crater taken by Surveyor 7. The hills on the center horizon are about eight miles away from the spacecraft. Since the landing site survey for the Apollo missions had been completed by the previous Surveyors, the landing site for Surveyor 7 was selected more for its scientific interest. Surveyor 7, in addition to taking thousands of images and gathering a wide variety of surface data, performed star surveys, took pictures of Earth, and tested laser-pointing techniques by detecting laser beams from Earth.

Gene Giberson was the JPL project manager. The Surveyor spacecraft were built for JPL by Hughes Aircraft Company. This was the first time that JPL (as a NASA Center) used a contractor to build a spacecraft. JPL and NASA learned many hard lessons about contracting with industry on this project. Surveyor experienced significant delays and cost overruns, but this learning process contributed to the success of future partnerships between JPL and industry.

Results

The Surveyors acquired almost 90,000 images from five sites and helped select landing spots for the Apollo missions. The next landings on the Moon will be the Apollo missions that would deliver astronauts safely to the Moon's surface.

However, once again the Soviet space program had beaten the American program, by landing Luna 9 (shown to the right) on the Moon on February 3, 1966, three months earlier than Surveyor 1.

Luna 9 was the first survivable landing of a human-made object on a body beyond Earth. It was Russia's 12th attempt to soft-land on the Moon and it was a remarkable success. The spacecraft performed flawlessly, touching down in the Ocean of Storms. It sent back the first images ever taken from the surface of another planetary body.

Mariner 5 – Venus Flyby

Mariner 5 was NASA's second mission to Venus. Dan Schneiderman was the manager.

This spacecraft was originally a backup to Mariner 4. When Mariner 4 completed its mission successfully, the backup was rechristened Mariner 5 and re-outfitted for a flyby of Venus. Mariner 5 carried a complement of experiments to probe Venus's atmosphere with radio waves, scan its brightness in ultraviolet light, and sample the solar particles and magnetic field fluctuations above the planet. Instruments came from JPL, Stanford University, MIT, the University of Iowa, and the University of Colorado.

Mariner 5 launched from Cape Canaveral, in June 1967. It flew within 2,500 miles of Venus on October 19, 1967. Mariner 5's flight path following its Venus encounter brought it closer to the Sun than any previous probe.

Mission operations for the October 1967 Mariner 5 fly-by of Venus are low-tech by today's standards. This image shows the technology in use at the time. Mission status data was located on boards around the room and numbers were changed by hand. "State-of-the-art" telephones, mechanical calculators, telex printers and cameras can be seen around the room.

Consequences

Mariner 5 studied the atmosphere of Venus, discovering its composition of 85-99% carbon dioxide. This discovery raised concerns about the effect of carbon dioxide on Earth's environment, a scientific issue that has been studied ever since.

A New Era at JPL

By 1967, JPL had pretty much worked out how to conduct first-of-a-kind missions. The management structure, planning process and basic life cycle was in place and worked quite well. Systems engineering methods were refined, including the "progressive freeze" approach to requirements and design information, with rigorous change control imposed on "frozen" requirements and designs. Engineering models (i.e., very accurate prototypes) started to be used. Testing was improved, as well.

JPL enjoyed nearly continuous technical success with its missions for the next twenty three years, except for (rare) failures of launch vehicles. If Cape Canaveral (later named Kennedy Space Center) could launch them, the mission would succeed.

One feature of this new era (1967 to 1990) was the use of a fairly standard spacecraft design, which evolved in capability from one mission to the next. Hard-won lessons from Ranger transferred over to the Mariner series of spacecraft, and beyond Mariner, this spacecraft line expanded to include most of the planetary missions of the 1970s and 1980s.

A subtle but important transition occurred in the matter of mission objectives.

The original objective was to gain technical knowledge at the fastest possible pace, in order to compete with the surprisingly capable Soviet space program. Science was part of each mission, but science had to fit within the limited capabilities and time frames of the early missions. The "space race" objective would remain important into the 1970s, but the importance of scientific achievement would increase from this point forward, until science became the dominant theme for nearly all JPL missions. The really big breakthroughs would primarily come from major new developments in scientific instruments.

Mariner 6 and 7 – The Second Mars Flyby

Mariner 6 and 7 was the second and final dual flyby of Mars. Bud Schurmeier was the Project Manager. The spacecraft was built at JPL, and instruments came from Caltech, UC Berkeley, and the University of Colorado.

Mariner 6 was launched on February 24, 1969. Mariner 7 was launched on March 27, 1969. Mariner 6 flew over the Martian equator on July 31, 1969, and Mariner 7 flew over the south polar region of Mars on August 5, 1969. The two spacecraft returned a combined total of 143 approach pictures of the planet and 55 close-up pictures. Their approach pictures showed that the dark features on the surface long seen from Earth were not canals. This was the first Mariner that included reprogrammable on-board computers, and Mariner 7 was reprogrammed to follow up on intriguing information from Mariner 6.

The "Happening"

Communications with Mariner 7 were lost as it approached Mars. Months later, it was determined that a battery had exploded, expelling gas that made the spacecraft rotate. Mariner 7 had lost the alignment it needed for solar power and regular communications. Mission operators at JPL scrambled to stabilize the spacecraft, just in time to save the Mariner 7. They then used the science cameras to recover the precise alignment needed for its approach to Mars, and the mission was successfully completed.

Wide-angle images of Mars were laid in place on a globe already containing an indistinct, earth-based view of Mars (see picture to the right).

The Mariner 6 pictures make two horizontal rows above; the Mariner 7 pictures extend from center to bottom right and across the south polar cap.

The Visual Imaging Investigation for Mariner 6 and 7 used two cameras on each spacecraft, for both broad coverage and high resolution.

Camera A had a wide-angle lens that showed large areas of the planet, 1000 x 1000 kilometers and details as small as 3 kilometers during the near encounter.

Camera B had a telephoto lens that showed 100 x 100 kilometer areas and details as small as 300 meters. The cameras operated alternately, with each one taking a picture every 84 seconds. An image from the high resolution camera is shown below.

A Back Story

Computer technology was both primitive and scarce in the 1960s. During mission development in 1967, the trajectory design model shown to the right allowed Mariner Mars mission planners to calculate the orientation of the planet, the expected path of the Mariner 6 and 7 spacecraft, and the window of opportunity for the instruments and TV cameras to operate during the flyby. Today, computer generated plots and computer animations are used to present the same information.

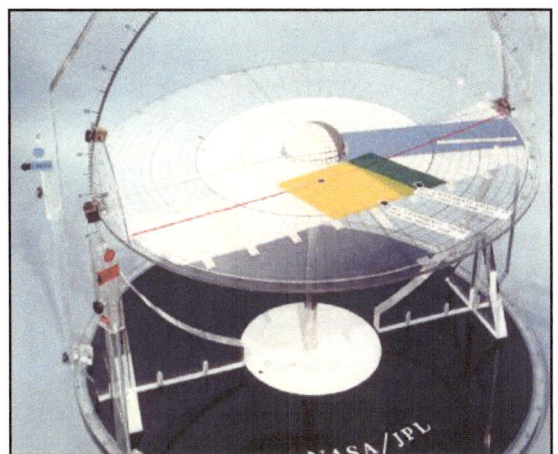

Mariner 8, 9 – the First Mars Orbiters

Mariner 8 and 9 were the final pair of Mars missions in the Mariner series, and the first to be designed as orbiters. Bruce Murray and Carl Sagan were scientists who worked on this mission. Dan Schniederman was the Project Manager. These were the last Mariner spacecraft built at JPL. Instruments came from Caltech, Santa Barbara Research Center (radiometer), JPL, Electro-Optical Systems (TV camera), Texas Instruments (spectrometer), University of Iowa, and the University of Colorado

Mariner 8 failed during launch on May 8, 1971, but Mariner 9 launched successfully on May 30, 1971, and achieved Mars orbit in September, 1971. This picture shows Mariner 8 being assembled.

This mission inherited the capability to reprogram the computers while in space (The central computer had an onboard memory of 512 words). Upon arrival, scientists observed a great dust storm which obscured the whole globe of Mars. Ground controllers sent new commands to the spacecraft to wait until the storm had abated and the surface was clearly visible before compiling its global mosaic of high-quality images of the Martian surface. The storm persisted for a month, but after the dust cleared, Mariner 9 was able to proceed with its mission.

The photo to the right shows the Mariner Mars spacecraft.

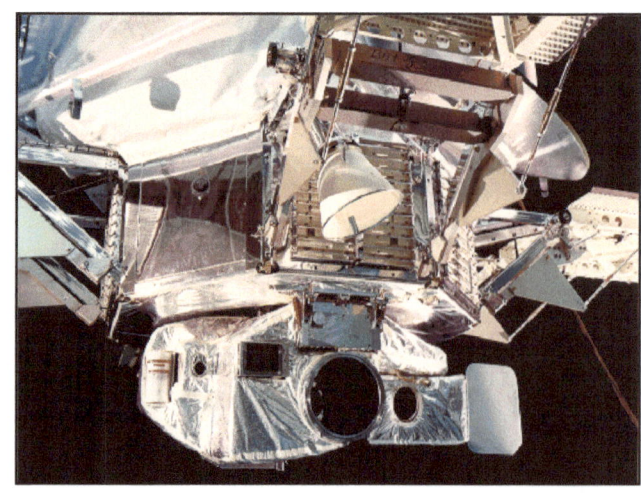

From left to right on the bottom is:
- The white cylinder of the infrared radiometer
- The wide-angle TV camera
- The ultraviolet spectrometer
- The narrow angle TV camera and
- The infrared spectrometer.

What is a Spectrometer?

This is an instrument that measures how much light of specific wavelengths is absorbed by an object. Typically it will plot a graph of absorption versus wavelength or frequency. The patterns produced are used to identify the substances present in the object. Spectrometers can operate using ultraviolet, infrared, visible, or other wavelengths of the electromagnetic spectrum. It is a cousin to the Radiometer, which measures the wavelengths of light emitted by an object.

The Impact of Mariner 9

Mariner 9 uncovered a very different planet than expected -- one that boasted gigantic volcanoes and a grand canyon stretching 3,000 miles across its surface. More surprisingly, the relics of ancient riverbeds were carved in the landscape of this seemingly dry and dusty planet. Mariner 9 exceeded all requirements by photo-mapping 100 percent of the planet's surface. The spacecraft also provided the first close up pictures of the two small, irregular Martian moons: Phobos and Deimos. Mariner 9 completed its final transmission on October 27, 1972.

Phobos is the larger of the two Martian moons and is dominated by three large craters. The largest crater is 10 kilometers wide, which is almost half of the average diameter of Phobos. Another interesting feature about Phobos is the duration of its orbit. It revolves around Mars three times during one Martian day, which is an astounding rate in this Solar system.

A Comment on the Space Race

Eleven Soviet missions to Mars up through 1971 had produced only 2 successes: Mars 2 and Mars 3. The Soviets tried to beat Mariner 8 by launching Cosmos 419 on May 10 (nearly the

same day as Mariner 8), but their stage four rocket failed, and it never left Earth orbit. The Soviets then launched the Mars 2 and Mars 3 (see picture to right) missions on 19 May and 28 May, respectively. Both went into Mars orbit in December, 1971. Mariner 9 achieved its Mars orbit in September, 1971 – a very narrow win for the American program.

At this point, there was relative parity between the Soviet deep space program and the American program. However, after this point in history, the American program pulled ahead, by a very wide margin. The Soviet program continued to move forward, but by the end of the 1980s the "space race" was largely replaced by cooperative efforts.

Mariner 10, the Mission to Venus and Mercury

The final Mariner mission was the first to go to Mercury, the first to employ a "gravity assist", and the first demonstration of a "solar sail". Gene Giberson managed the mission. The spacecraft was built by Boeing. Instruments were provided by Caltech, MIT, Goddard Spaceflight Center, the Kitt Peak national Observatory, JPL, and the University of Chicago.

It launched on November 3, 1973, with a Venus flyby on February 5, 1974 and Mercury flybys on March 29, 1974; September 20, 1974; and March 16, 1975.

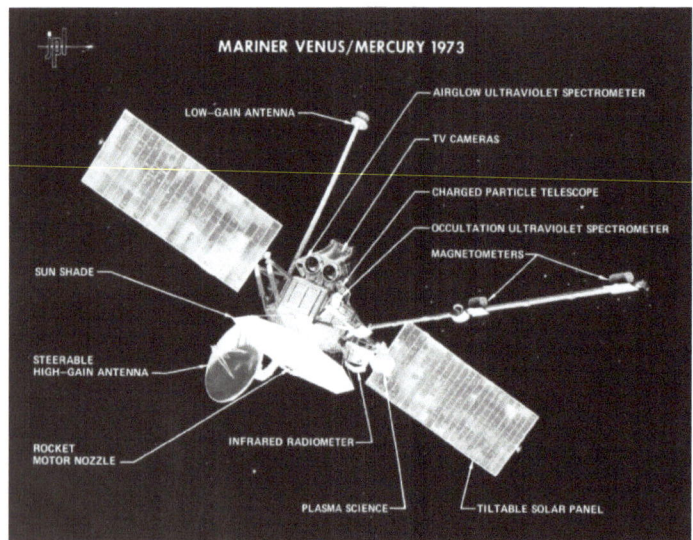

The image to the left shows the spacecraft and all of its instruments.

On February 5, 1974, Mariner 10 made its first flyby of Venus and discovered evidence of rotating clouds, taking some 4,000 photos of Venus.

Mariner 10 pioneered the 'Gravity Assist', using Venus to assist its approach to Mercury, and Mercury's gravity was also used to change its trajectory -- allowing it to fly by Mercury three times at roughly 6 month intervals. This craft was also the first to use the solar wind as a means of locomotion; when the probe's thruster fuel ran low, engineers used the solar panels as sails to make final course corrections.

On March 29 it flew within 420 miles of Mercury and using gravity-assist techniques, returned twice during the following year. The spacecraft returned about 4,000 Mercury images, mapping 45% of the planet surface. It discovered that Mercury had a thin atmosphere and a magnetic field.

The image to the right was taken during the first encounter with Mercury.

The Kuiper crater was named in memory of Gerard P. Kuiper, a pioneer in planetary astronomy and member of the Mariner 10 TV team.

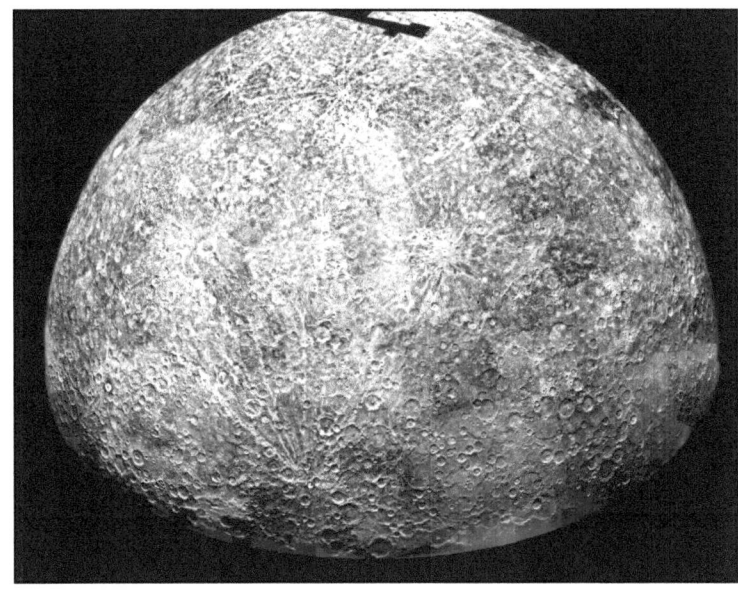

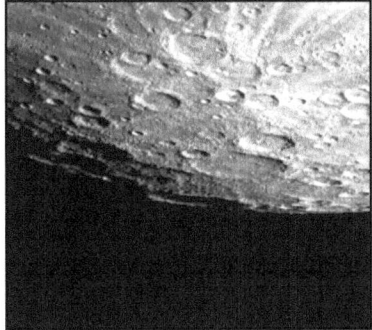

Mercury was encountered a second time at a distance of 29,870 miles. To the left is a picture of Mercury's South pole.

There was a third Mercury fly by, but Mercury's slow rotation left the other half in the dark when Mariner 10 returned. Passing 203 miles above the planet, its mapping mission was complete. By mapping about half the planet, the mission showed a cratered surface and a faint, mostly helium atmosphere.

Results
Mariner 10 was the only encounter with the planet Mercury for 25 years, until the Messenger flew past Mercury in January, 2008.

The radio science and ultraviolet experiments confirmed that Mercury, like the Earth's Moon, lacks an atmosphere.

Mercury rotates three times every two times it circles the sun. Thus, a Mercury day is two thirds as long as a Mercury year and there is plenty of time for the hot side to heat up and the cool side to freeze. The infrared experiment determined that the surface temperatures range from -297 degrees F on the night side of Mercury to +369 decrees F on the day side.

Surprisingly, Mercury has a magnetic field. It was believed that the low rate of rotation of Mercury would make a magnetic field unlikely, but it was found to exist anyway.

The surface of Mercury has been subjected to a similar meteoric bombardment as Mars and the Moon, which argues against the theory that the Martian and Lunar bombardments are due to random strikes from objects in the asteroid belt. The distribution of such strikes would vary with distance from the asteroid belt itself, and the actual distributions are quite similar.

A Back Story
Michael Minovitch showed that spacecraft trajectories could be designed to gain or lose velocity by travelling close to a planet orbiting the sun. This technique was developed in the early 1960s when he was a UCLA graduate student working summers at JPL. JPL has always employed students and graduate students as a means of bringing in fresh ideas from academia to the laboratory. In 2008 JPL had 270 student employees in the summer and about 160 part time student employees in the fall.

Viking Orbiter

NASA's Viking Project included two identical spacecraft, show below. Each consisted of a lander and an orbiter. Their objective was to be the first mission to land a spacecraft safely on the surface of Mars.

This image shows one of the two identical Viking spacecraft, each of which included an orbiter and a lander. The lander was encapsulated in both a bioshield and an aeroshield, which can be seen at the bottom of the orbiter bus. The image also shows the black high gain antenna, solar panels extended for flight, and the covered fuel and oxidizer tanks.

JPL was responsible for building the orbiters, providing the Mission Control and Computing Center, and for tracking and data acquisition using the Deep Space Network. Cameras were built by the Ball Brothers Research Corporation for Michael Carr of the U.S. Geological Survey; an infrared thermal mapper was built by the Santa Barbara Research Center for Hugh Keiffer of UCLA; and a water vapor mapper by JPL. Henry Norris and K. S. Watkins were the JPL Project Managers.

The landers were developed and built by Martin-Marietta Corporation under contract to Langley Research Center, which had overall project management responsibility.

The orbiters were based on Mariner heritage, but were much larger (5,125 pounds each), with more propulsion, power, and telecommunications capability. They both had a pair of advanced, all-purpose on-board computers with a memory of 4,096 words. They would relay communications from the landers to Earth, using 8 track tape digital recorders for intermediate data storage. The orbiters were designed to last until 90 days after the Viking landing. They actually continued working for 2 (Viking 1) and 4 (Viking 2) years.

The Viking 1 spacecraft was launched on August 20, 1975 and Viking 2 was launched a few weeks later, on September 9. Viking 1 and 2 reached Mars in June and August, 1976.

This mosaic of Mars images was taken by Viking Orbiter 1 over a period of 6 days in August 1976. It shows channels and craters west of Chryse Planitia, the site where Viking Lander 1 touched down. The channels suggest a massive flood of water from Lunae Planum, flowing across this cratered terrain. In some cases, the channels cut through craters. In others, the craters clearly happened later than the flood and are superimposed on the channels.

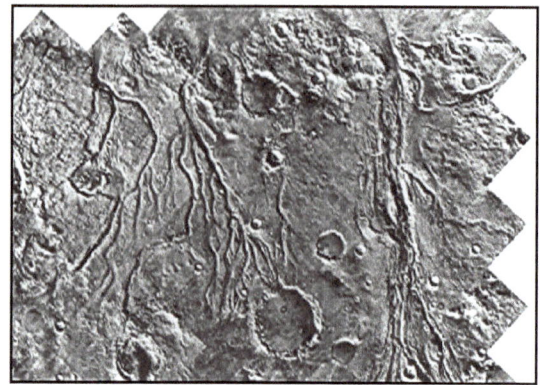

The orbiters also successfully delivered their landers, as shown in the picture below.

The orbiter cameras then mapped 97% of the Martian surface, sending back a total of 52,000 images. Infrared thermal mappers and atmospheric water detectors also returned data throughout the mission. Viking Orbiter 1 functioned until July 25, 1978, while Viking Orbiter 2 orbited Mars 1,489 times, concluding the mission on August 7, 1980.

Results

The image to the right shows a massive dust storm, blowing across the surface of Mars. This large disturbance soon grew into the first global dust storm observed by the Viking Orbiters.

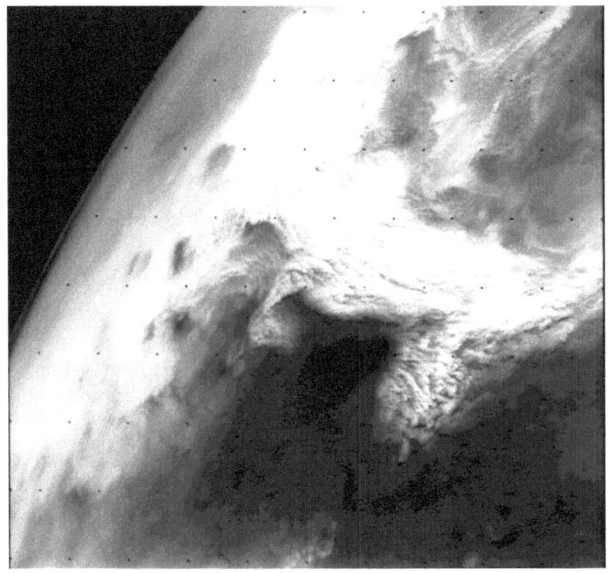

The image below is the famous 'face on Mars'. It was taken on July 25, 1976 while searching for a landing site for the Viking 2 Lander. The speckled appearance of the image is due to

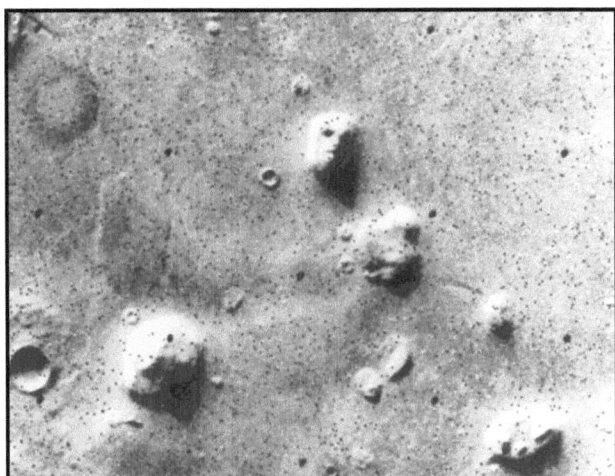

missing data, called bit errors, caused by problems in transmission of the photographic data from Mars to Earth. Bit errors comprise part of one of the 'eyes' and 'nostrils' on the eroded rock that resembles a human face near the center of the image. Shadows in the rock formation give the illusion of a nose and mouth. Planetary geologists attribute the origin of the formation to purely natural processes. The feature is one mile across, with the sun angle at approximately 20 degrees.

The orbiters determined the composition of the Martian atmosphere, including isotopes of various gases. From this data, scientists have decided that the atmosphere of Mars was much denser in the past than it is today. Also, atmospheric density varies by 30% over the course of the Martian year, due to carbon dioxide condensing in the winter in the Polar Regions.

The images of the Martian surface revealed a complex history of Mars, completely different from our understanding of any other planet.

Voyager – The Grand Tour

In 1966, JPL had proposed to NASA a "Grand Tour" of the outer planets, including Pluto. Each planet's gravity would accelerate the spacecraft to the next target. To travel such vast distances, the spacecraft would be powered by radioisotope thermal generators. The Grand Tour would last 12 years. NASA approved the Grand Tour in 1971, with Bud Schurmeier as the project manager, then cancelled it in 1972. Within a few weeks of the cancellation, Schurmeier's team came back with a plan to build a Mariner-class spacecraft, designed for a more modest four-year mission to Jupiter and Saturn. It would push the envelope on the Mariner spacecraft for about one quarter the cost of the Grand Tour, but with no formal plan to go on to Uranus and Neptune. The mission was originally called Mariner-Jupiter-Saturn 1977. The JPL team, however, wanted to extend the Tour and decided not to limit the spacecraft's capabilities to the approved Jupiter/Saturn mission. If the vehicles survived Saturn, they could continue to Uranus and Neptune (but the delay in launch had ruled out Pluto).

MJS 1977 was renamed "Voyager" shortly after John Casani became project manager. Casani thought MJS was a terrible name for a robotic explorer and advocated a more distinctive title. One of the Voyager spacecraft is shown to the right, being prepared for testing in the solar thermal vacuum chamber in JPL's Environmental Test Laboratory.

Both spacecraft carry a "golden record" to greet any form of life that might be encountered. The contents were selected for NASA by a committee chaired by Carl Sagan of Cornell University. It has 115 images and a variety of natural sounds. To this they added musical selections from different cultures and eras, and spoken greetings from Earth-people in fifty-five languages.

Both spacecraft were loaded with science instruments from the University of Arizona (cameras); Caltech (cosmic-ray detectors); GSFC (infrared spectrometers, radiometers, and magnetometers); Johns Hopkins University (charged-particle detectors); JPL (a photopolarimeter); University of Colorado (radio astronomy instruments); MIT (a plasma experiment); TRW (a plasma wave experiment); and the Kitt Peak National Observatory (UV spectrometers).

There is a distinguished roster of project managers on the Voyager mission: 'Bud' Schurmeier (1970 to 1976); John Casani (1976-1977); Robert Parks (1978-1979); Raymond Heacock (1979-1981); Esker Davis (1981-1982); Richard Laeser (1982-1986); Norm Haynes (1987-1989); George Textor (1989-1997); Ed B. Massey (1998 - present).

This was Schurmeirer's last flight project – his final years were spent in the JPL program for developing renewable energy – where I first met him. It was the first project management job for John Casani, who will be mentioned many times from this point onward.

But there has been only one project scientist, Ed Stone, who started in 1972.

Voyager 2 launched first, on August 20, 1977. Voyager 1 followed on September 5. Because of the constantly changing relationships among the planets, Voyager 1 arrived at Jupiter first, in March 1979. Voyager 2 made its closest pass on July 9, 1979.

The image to the left is the first image of Jupiter. The image to the right is a close up of Jupiter's Great Red Spot, seen by Voyager 1.

Voyager also studied the moons of each planet it encountered. These turned out to be wonderfully diverse.

These images are Europa and Callisto (far right).

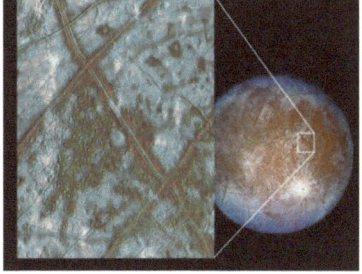

Voyager 1 reached Saturn in November 1979, with Voyager 2 following in August 1981. At Saturn, the two spacecraft discovered new rings, imaged moons that could not be seen from Earth and examined Saturn's atmosphere.

This true color picture was assembled from Voyager 2 Saturn images obtained Aug. 4 from a distance of 13 million miles on the spacecraft's approach trajectory.

Three of Saturn's icy moons are evident in this image. They are, in order of distance from the planet: Tethys, 652 miles in diameter; Dione, 696 miles; and Rhea, 951 miles. The shadow of Tethys appears on Saturn's southern hemisphere. A fourth satellite, Mimas, is less evident, appearing as a bright spot a quarter-inch in from the planet's limb about half an inch above Tethys; the shadow of Mimas appears on the planet about three-quarters of an inch directly above that of Tethys.

The pastel and yellow hues on the planet reveal many contrasting bright and darker bands in both hemispheres of Saturn's weather system.

Voyager 1's survey of Saturn's largest moon, Titan was somewhat disappointing. As seen to the right, this moon's atmosphere is completely opaque, and Voyager's images revealed only a fuzzy orange ball. But the atmosphere was denser than Earth's, composed primarily of molecular nitrogen, with the orange color apparently coming from methane. Titan was so cold that methane seemed to act like water on Earth, cycling between solid and liquid forms on the surface, and liquid and gaseous forms in the atmosphere.

Much more visible was Saturn's moon, Dione. Note the many impact craters. This photo was taken on November 12, 1980, at a range of 129,000 km

Voyager 2 made its flyby of Uranus on January 24, 1986. Scientists were rewarded at Uranus with the discovery of two new rings and ten moons. This is a view of Uranus taken by Voyager 2. This image was taken through three color filters and recombined to produce the color image.

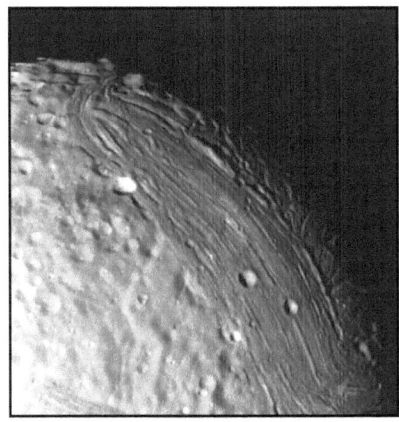

Images of Miranda, the closest large moon to the planet, revealed an exceedingly strange surface. Parts of it were cratered like Earth's moon, while other parts looked as if they had been plowed with gigantic farm tools. The mission scientists have no explanation for this bizarre landscape. The view to the left was acquired by Voyager 2 on Jan. 24, 1986, around its close approach to the Uranian moon.

Uranus, which appeared as an almost featureless green ball, turned out to have an unusual magnetic field that was offset from the planet's core by thousands of kilometers. This was possible only if Uranus had a very non-Earthlike internal structure, again leaving Voyager scientists without a compelling explanation for the planet's strangeness.

It took Voyager 2 three-and-one-half years to reach the last planet on its Grand Tour. The spacecraft passed a bare 4,500 kilometers (about 2,800 miles) above Neptune on August 25, 1989. As Voyager 2 approached, its images revealed that Neptune had an Earth-sized "great dark spot" like Jupiter's red spot, immersed in a similar, if less colorful, set of atmospheric bands.

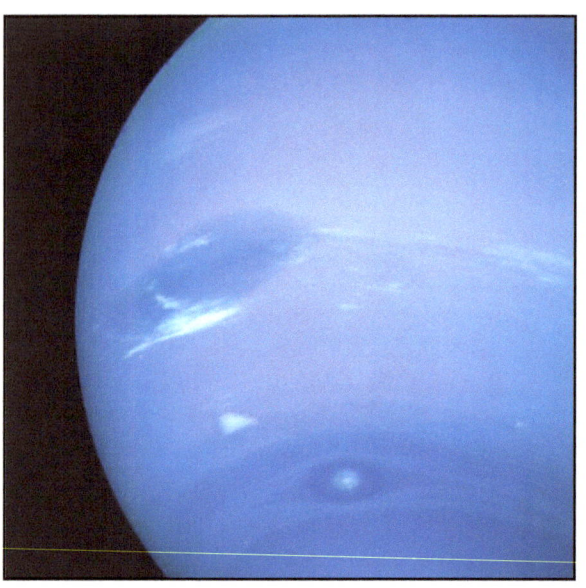

The image to the left shows the "Great Black Spot" on Neptune, taken by Voyager 2 on Jan. 24, 1986.

The Voyager data also confirmed the existence of Neptunian rings. Mission scientists used these findings to determine that like Uranus, Neptune's magnetic field was offset far from the core. Unlike Uranus, Neptune had a source of internal heat, leading to the highest velocity winds of any planet.

Finally, Voyager 2 imaged active nitrogen-ice volcanoes on Triton, the largest of Neptune's moons.

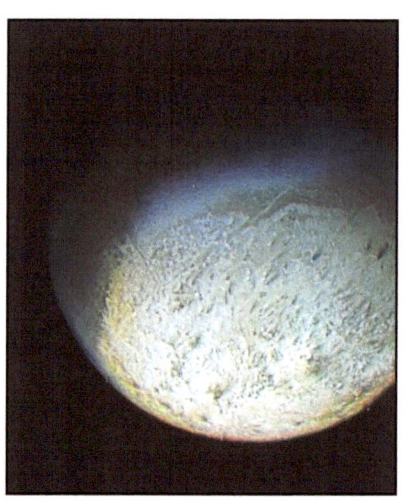

After Voyager 2's Neptune encounter, NASA renamed the Voyager project the Voyager Interstellar Mission. Most of the instruments on both spacecraft were permanently deactivated, but solar wind-related instruments kept operating to monitor the sun's waning influence. In December 2004, Voyager project scientist Ed Stone announced that Voyager 1 had passed through the 'termination shock,' where the solar wind abruptly slows to subsonic speeds. It was 94.01 astronomical units from the sun. (Earth is one astronomical unit from the sun.)

In August 2006, Voyager 1 reached 100 astronomical units from the sun, or about 9.3 billion miles from the sun. Voyager 2 was about 7 billion miles from the sun. Both craft are healthy and continue to send data back to Earth. In the next ten years, scientists expect the Voyagers to cross the heliopause, the edge of the bubble created by the solar wind, and become the first craft to reach interstellar space.

Summary
Between them, Voyager 1 and 2 explored all the giant planets of our outer solar system, Jupiter, Saturn, Uranus and Neptune. They also explored 48 of the moons and the unique systems of rings and magnetic fields these giant planets possess.

Ed Stone summarized the whole experience: "Voyager has become the icon of interplanetary missions. It saw more new worlds for the first time than any mission ever has." Voyager was an incredible journey of discovery. One of the things that made it so exciting was that all these planets and moons could have been the same. But what we found was that "nature was incredibly inventive in the way it used the basic laws of nature to create these diverse, distinct worlds."

A Note on Bruce Murray
There was a change in JPL management a year before the Voyagers were launched. Bill Pickering retired, and Bruce Murray became JPL's new Director in April, 1976.

As NASA's planetary science budget shrank, Bruce Murray expanded the lab's research horizons to preserve its technological capabilities.

JPL took on new tasks in Earth science and astronomy, while continuing its leading role in planetary exploration. The first new Earth science mission would revolutionize the fields of oceanography and remote sensing.

Seasat – Revolutionizing Oceanography

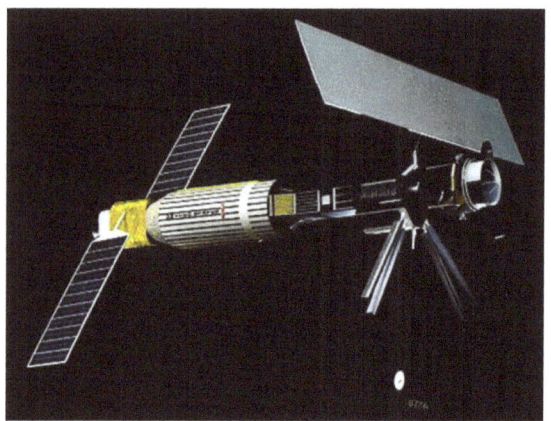

JPL engineers and scientists realized that sensors developed for interplanetary missions could also study Earth.

Seasat launched on June 26, 1978. It was the world's first dedicated oceanographic satellite, testing revolutionary new sensors. It carried four complementary microwave experiments: a radar altimeter to precisely measure spacecraft altitude above the ocean surface; a scatterometer to measure wind speed and direction over the ocean; a radiometer to measure ocean surface temperature, atmospheric water vapor content, rain rate, and ice coverage; and a synthetic aperture radar to image the ocean surface, polar ice caps, and coastal regions. Thousands of papers have been produced using Seasat data or data from instruments derived from the Seasat experience.

Gene Giberson was the Project Manager. Dr. James Dunne coordinated experiments that came from JPL (radars), the University of Texas, the U.S. Geological Survey, the University of New York, the National Environmental Satellite Service, and the Atlantic Oceanographic Laboratory. Lockheed Missiles and Space Company built the spacecraft.

What is a Scatterometer?

Scatterometers use radar or microwave energy that is beamed down to a target, and the reflected energy that returns to the instrument is analyzed to measure waves, and from this it is possible to calculate near-surface winds over the ocean.

The image to the right shows Seasat's Synthetic Aperture Radar (SAR) antenna being unfolded from launch configuration to its full size of 7 feet by 35 feet. Seasat pioneered the use of SAR remote sensing of Earth's oceans.

The Mission

For over three months Seasat observed sea-surface winds and temperatures, wave heights, internal waves, atmospheric water, sea ice features and ocean topography. These were the first global sets of data of this type ever recorded.

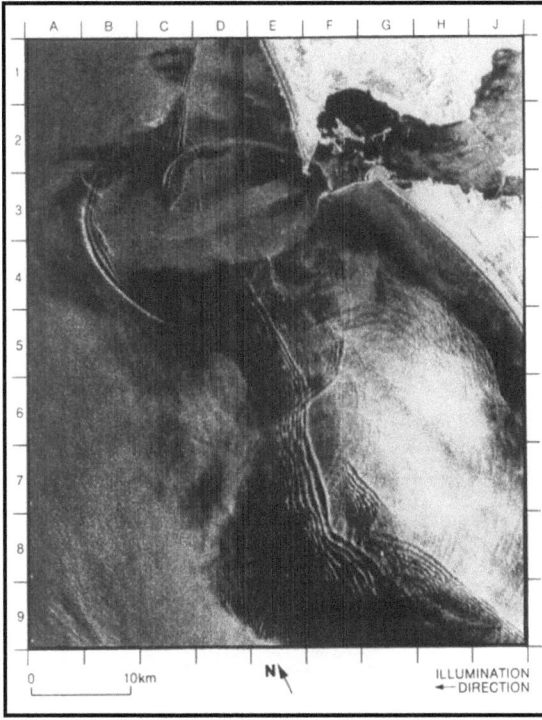

Oceanographers wanted measurements from space to help them understand how Earth's oceans circulate. They have measured ocean currents from ships for centuries (this was how Benjamin Franklin discovered a portion of the Gulf Stream), but ships only measure a tiny portion of the ocean at a time. Satellites can see large swaths of Earth at once.

Seasat gave us our first global view of ocean circulation, waves and winds. It provided new insights into the links between the ocean and atmosphere that drive our climate. The image to the left shows wave refraction at the mouth of the Columbia River.

Seasat failed after only 105 days in orbit. NASA's investigation of the failure determined that there was a bad power coupling in the satellite's solar panels.

Going from Failure to Success

Seasat was supposed to last for at least 1 year, but sent data to Earth for only 3 months. Initially the mission was judged a failure, due to its short duration.

But the volume and quality of the data obtained from Seasat enabled scientists to greatly advance their understanding of oceanic and weather phenomena. The scientific analysis of data continued for years after the mission, and Seasat was re-evaluated as a major success, due to the value of the science data returned.

Results

Seasat changed the field of oceanographic research by adding a new methodology – space-based observation—to a field that has been around for centuries. "Seasat served to vault Earth science to where it is today, advancing the study of such diverse disciplines as land and sea surface topography, ice sheet and land movement, and sea surface winds," said Dr. Frank Carsey, JPL research scientist. "It greatly advanced our understanding of the El Nino and La Nina climate phenomena. It's astonishing to think such a short mission could have such a tremendous impact."

For the first time, an entire section of the ocean could be seen all at once. Seasat's altimeter mapped ocean topography, allowing scientists to determine ocean circulation and heat storage.

The data also revealed new information about Earth's gravity field and the topography of the ocean floor. Seasat also revealed information about ocean pollution, as seen in the image to the right where the black areas show oil slicks off the coast of California.

Because of Seasat, advanced ocean altimeters on JPL's Topex/Poseidon and Jason missions have been making precise measurements of sea surface height used to study climate phenomena such as El Nino and La Nina. Ocean altimetry has since become part of weather and climate models, ship routing, marine mammal studies, fisheries management and offshore operations.

Seasat's scatterometer gave us our first real-time global map of the speed and direction of ocean winds, which drive waves and currents and are the major link between the ocean and atmosphere.

Seasat's oceanographic mission also studied sea ice and its role in controlling Earth's climate. Its synthetic aperture radar provided the first high-resolution images of sea ice, measuring its movement, deformation, age and thickness. Today, synthetic aperture radar and scatterometers are both used to monitor Earth's ice from space.

Beyond the oceans, Seasat's synthetic aperture radar provided spectacular images of Earth's land surfaces and geology. Seasat data was used to pioneer radar interferometry, which can pinpoint land surface changes such as those created by earthquakes, and to measure land surface topography.

"Seasat had a major impact on future mission planning at NASA and elsewhere," said Tony Spear, the Seasat payload manager. "Its prototype radars and altimeter were precursors for many of today's more powerful Earth observation satellites."

Solar Mesosphere Explorer – Measuring Earth's Ozone

The Solar Mesosphere Explorer (SME) mission was developed to investigate the creation and destruction of ozone in Earth's upper atmosphere. SME had 5 instruments that simultaneously measure Ozone in the Mesosphere; water vapor; and incoming solar radiation.

Definition of Mesosphere – the layer of Earth's atmosphere that is above the stratosphere. This is about 20 to 50 miles above the Earth's surface.

SME was managed by JPL. The instrument and spacecraft were built by Ball Space Systems and operated by the University of Colorado.

SME launched from Vandenberg Air Force Base, California on October 6, 1981, and sent data to Earth until April 4, 1989. The spacecraft reentered Earth's atmosphere on March 5, 1991. The white are in the image below shows the ozone hole in September 1987.

Significance

The ozone layer is a thin band in Earth's upper atmosphere. It blocks out the Sun's harmful ultraviolet (UV) rays. If it gets too thin, the harmful UV rays can damage crops, wild animals, and our skin. SME monitored the ozone hole, which is a zone of greatly depleted ozone that expands and contracts over the south polar region. This discovery led to a series of missions to monitor the ozone layer.

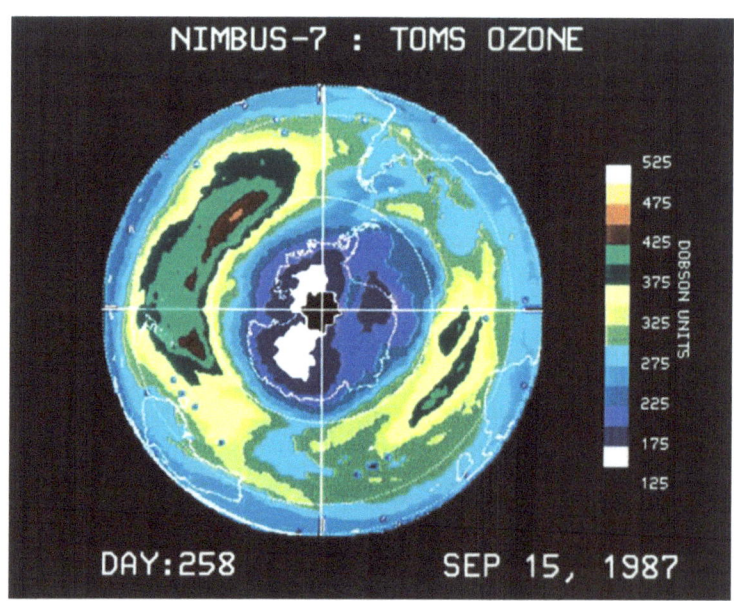

Results

In 1987, diplomats from most of the world's nations met in Montreal, Canada, to negotiate a mandatory reduction in chlorofluorocarbon production in response to NASA's data. In 1992, this response was further tightened to ban production of these chemicals entirely. JPL's Microwave Limb Sounder on the Aura satellite still monitors stratospheric health.

Shuttle Imaging Radar – Exploring Earth

Shuttle Imaging Radar are four different but related projects that flew synthetic aperture radar (see image below) on the Space Shuttle in order to create better maps of the Earth.

Seasat had established that images could be taken of Earth from orbit using radar pulses rather than optical light as the illumination. Imaging radar can "see" through desert sands, for example, to detect the remnants of ancient riverbeds. A series of imaging radar missions were therefore flown on NASA's Space Shuttle over the next 20 years.

The first mission, called Shuttle Imaging Radar-A, was carried into space on STS-2, only the second mission flown by the then-new shuttle. This mission launched on November 12, 1981, and landed two days later. Charles Elachi, the future Director of JPL, was the principal investigator.

Three years later, a follow-up mission called Shuttle Imaging Radar-B was flown on shuttle mission STS-41G, which launched October 5, 1984, and landed seven days later.

After that project, a decade went by before imaging radar flew on the shuttle again.

For the next mission, JPL's Spaceborne Imaging Radar-C was combined with a German-Italian X-Band Synthetic Aperture Radar which used a higher-frequency radar than the American instrument. This package flew twice on the space shuttle, once on STS-59 from April 9 to 20, 1994, and the second time on STS-68 from September 30 to October 11, 1994 (STS-68 is shown to the right).

The Lost City of Ubar

The image to the right was used to locate the Biblical city of Ubar, which had been lost for nearly a thousand years. The lines are ancient roads, largely covered by sand, which were discovered by the SIR-C radar. Where they intersect, the lost city of Ubar was rediscovered by archeologists.

Michael Sander was the SIR-C Project Manager.

The four images shown here are all images of the Long Valley region of east-central California. They illustrate the steps required to produce three dimensional topographical maps. All data displayed in these images were acquired by the SIR-C aboard the space shuttle Endeavour during its two flights in April and October, 1994. The upper left shows radar data for an area that is 21 by 37 miles. The bright areas are hilly regions that contain exposed bedrock and pine forest. The darker

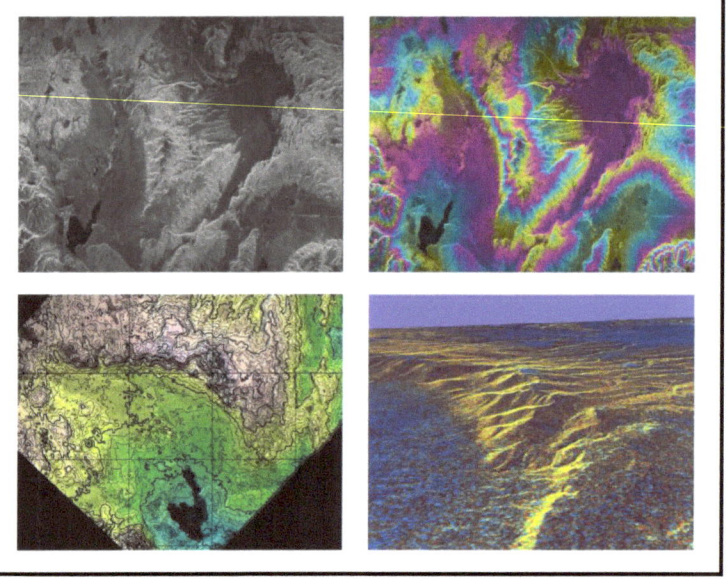

gray areas are the relatively smooth, sparsely vegetated valley floors. The dark irregular patch near the lower left is Lake Crowley. The curving ridge that runs across the center of the image from top to bottom is the northeast rim of the Long Valley Caldera, a remnant crater from a massive volcanic eruption that occurred about 750,000 years ago. The image in the upper right combines data from the April and October flights. The colors in this image are caused by elevation differences. The image in the lower left shows a topographic map, with thin black contour lines, which are spaced at 50-meter (164-foot) elevation intervals. Heavy contour lines show 250-meter intervals (820-foot). The image in the lower right is a three-dimensional perspective view of the northeast rim of the Long Valley caldera, looking toward the northwest. Combining topographic and radar image data allows scientists to examine relationships between geologic structures and landforms, and other properties of the land cover, such as soil type, vegetation distribution and water characteristics.

The instrument's mammoth radar antenna was then augmented with a second antenna that would allow it to map the height of features on Earth. Sponsored by the Defense Department's National Image Mapping Agency, this package flew under the name Shuttle Radar Topography Mission (SRTM) on STS-99 from February 11 to 22, in 2000. The major innovation of SRTM was that additional antennas were deployed on a 60-meter mast so the system could operate as a single-pass interferometer to efficiently and accurately collect elevation data. The 60-meter mast is the longest structure ever flown in space. Charley Yamarone was the Project Manager during implementation. Yunjin Kim was the Project Manager during the analysis phase of this project.

SRTM obtained more than 12 terabytes of raw data that was processed into research-quality digital elevation models for every continent. This is the most complete high-resolution digital topographic database of Earth. One example of how this is used is this view of the potential effects of storm surge flooding on Galveston and portions of south Houston from Hurricane Rita. This prediction came from the National Weather Service on September 22 at 4 p.m. Central Time. It shows the expected track center in black with the lighter shaded area indicating the range of potential tracks the storm could take. SRTM data showed the extent of low-lying terrain along the coast, with areas within 15 feet of sea level shown in red, and within 30 feet in yellow. These areas are more at risk for flooding and the destructive effects of storm surge and high waves.

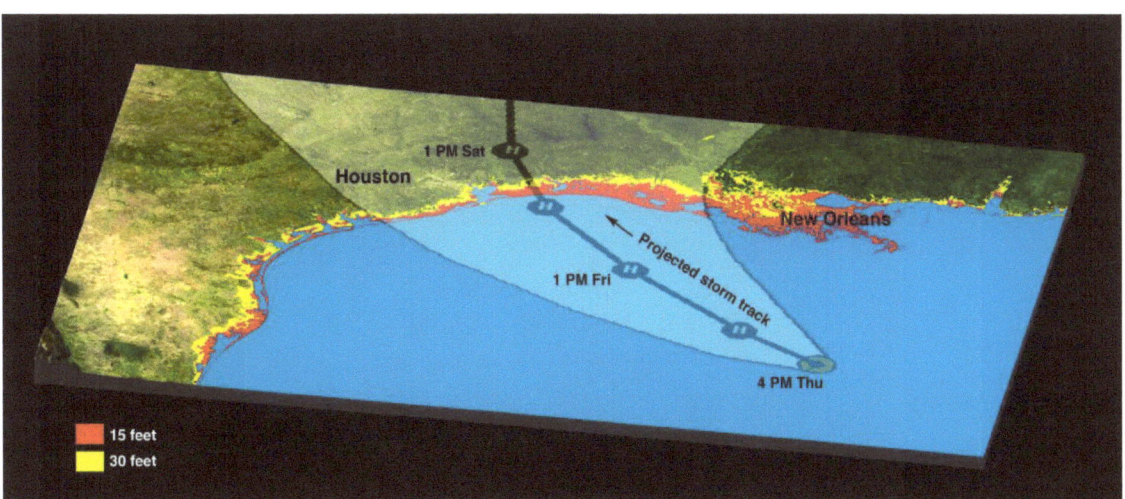

Another Note on JPL Management

In 1982, Bruce Murray retired as JPL Director.

His successor, Lew Allen, gained three missions for JPL. Two were planetary missions, Mars Observer and Magellan. The third was an Earth science mission, Topex/Poseidon.

Allen restructured JPL by increasing JPL's focus on space flight, and associated technology.

In the Allen era, JPL also became a significant builder of scientific instruments, many of which are described later in this book.

Infrared Astronomical Satellite (IRAS) – Surveying the Universe

The Infrared Astronomical Satellite (IRAS) was the first telescope that observed infrared images from an orbit above the interference of Earth's atmosphere.

Infrared telescopes require a special cooler to bring the temperature of the infrared detector down to a few degrees above absolute zero. This extremely low temperature made the detectors up to a thousand times more sensitive than any before them.

Gael Squibb was the Project Manager. The Netherlands provided the spacecraft, the United States contributed the launch, the infrared telescope and final data handling; the United Kingdom was in charge of satellite operations and preliminary data processing.

Launched on January 25, 1983, from California's Vandenberg Air Force Base, the satellite orbited Earth at an altitude of 563 miles. It operated for 10 months until its coolant was depleted as planned. The image below shows a mosaic of the universe in infrared. The bright band in the middle is our galaxy, the Milky Way.

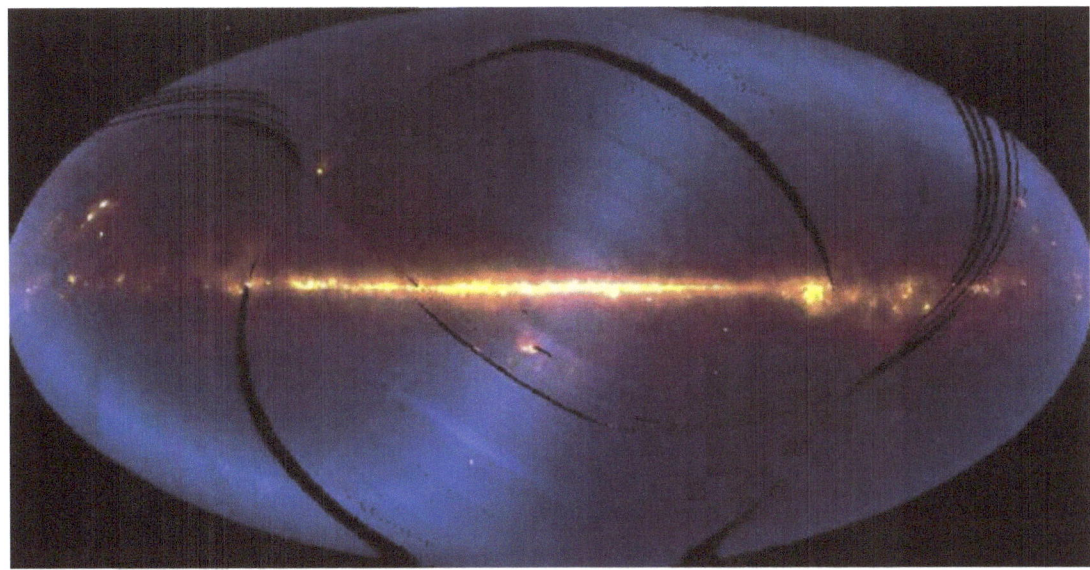

51

Results

IRAS completed a survey of 96 percent of the sky, locating and measuring more than 200,000 objects. This enabled astronomers to compile the first nearly complete atlas of the sky at infrared wavelengths.

IRAS enabled identification of very old stars, too cold to generate much visible light.

The satellite also detected huge masses of dust. It found that the bright star Vega was shining ten to twenty times brighter than it should have at long infrared wavelengths, a phenomenon known as "infrared excess". Additional observations showed that the source of the infrared excess was a ring of dusty material surrounding Vega, possibly the first evidence for a solar system in formation. No one had observed this before; to this point, astronomers could only assume that this phenomenon occurred.

 Two of the most significant findings were the discoveries of solid material around the stars Vega and Fomalhaut, located some 26 and 22 light-years from Earth, respectively. This strongly suggested the existence of planetary systems around other stars.

Atmospheric Trace Molecule Spectroscopy (ATMOS)

ATMOS was an infrared spectrometer designed to study the chemical composition of the atmosphere.

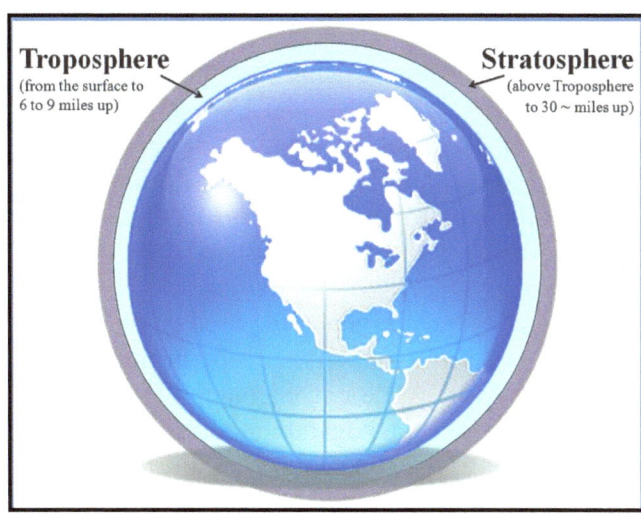

ATMOS has flown aboard the space shuttle four times, starting in 1985 producing simultaneous measurements of the vertical profiles of a variety of atmospheric gases from the upper troposphere to well above the stratosphere. These investigations provided fundamental data on the concentrations of greenhouse gases, chemicals that affect the ozone hole, and other components of the atmosphere that may be important to modeling climate change.

First flight: Spacelab 3   Launch: April 30, 1985

Second flight: STS-45/ATLAS 1   Launch: March 24, 1992

Third flight: STS-56/ATLAS 2   Launch: April 8, 1993

Fourth flight: STS-66/ATLAS 3   Launch: November 3, 1994

In the picture to the left, ATMOS is located to the right of the black sphere at the far side of the orbiter, and behind the instrument with the dish.

The Project Manager was Larry Simmons.

A Modest Contribution

I started at JPL in 1976, working as an analyst on renewable energy R&D projects through 1984 and information systems through 1990.

In 1988, I analyzed the performance of JPL's information system that solved navigation problems. At that time, JPL had tens of millions of lines of software code invested in a pair of Univac 1108 mainframe computers that dated back to the 1970s. These had limited memory (64,000 words) and very limited speed.

Solving a navigation problem required an overnight run, due to the size of these problems and the speed of those computers. This performance was acceptable for a flyby mission, which had months to prepare for an encounter. But it would only work as long as only one or two missions had critical navigation needs at a time. It was evident that this level of performance would be totally inadequate for a complex orbital mission, and these were coming up (e.g., Galileo, Mars Observer, and Magellan).

I was asked to confirm that a larger, more modern mainframe would be sufficient for JPL's future missions. My analysis showed that this strategy would not work, and that management should opt instead for distributed processing, where each mission bought and operated its own computers. (We called these "mid-frames" back in the late 80s.)

Management accepted my recommendation and invested over $40 million to move critical software to this new computing environment. They also upgraded JPL's networks to fiber optics at this time, to accommodate traffic between the "mid-frame" computers. This was before internet browsers changed the nature of computing, and before server technology became common place.

The new computing architecture has enabled JPL to maintain a very large number of spacecraft in flight operations in the 1990s and in this decade. Recently JPL has conducted mission operations on as many as 30 separate spacecraft concurrently.

Galileo – Exploring Jupiter

The birth of a flight project can be long and difficult.

The mission was originally proposed to start in 1977. The major challenge was to find a method of propulsion that could launch on the Space Shuttle (it was a mission requirement that Galileo launch from the Shuttle) and send a very large spacecraft all the way to Jupiter.

The original mission was cancelled in 1980.

The team that was proposing the Galileo mission continued to work, and came up with a concept of sending Galileo up in two sections on two separate Shuttle flights, and the two sections would be mated in low Earth orbit, and could then proceed to Jupiter.

This next version of Galileo was then cancelled in 1981.

The mission team continued to work, and came up with a mission that could launch in May of 1986. Galileo was built and was shipped to Kennedy Space Center in December, 1985. There it would wait for its launch opportunity in May, 1986.

But the Challenger tragedy occurred in January, 1986. The rocket that Galileo needed to get to Jupiter would never again be allowed onto the Shuttle, and Galileo was cancelled a third time.

The mission team went back to work to find a new way to make the mission work. They discovered a combination of gravity assists that had never been considered before – to go to Venus first (closer to the sun and further away from Jupiter), and then a pair of Earth flybys in order to build up the speed needed to make it to Jupiter. This approach reduced the energy needed to get to Jupiter by more than a factor of six. The Galileo mission was again proposed and was accepted.

Galileo launched on October 18, 1989 from the Kennedy Space Center. It was carried into Earth orbit in the cargo bay of Space Shuttle Atlantis. It was then propelled onto its interplanetary flight path by a two-stage solid-fuel motor called an Inertial Upper Stage. Galileo flew past Venus on February 10, 1990, and then twice past Earth, once on December 8, 1990, and again on December 8, 1992.

The spacecraft shown below had a unique feature: two sections were joined together by a spin bearing that works something like a lazy susan. Half of the spacecraft held pointable instruments such as cameras, and was held fixed in relation to space. The other half of the spacecraft contained instruments that measured magnetic fields and charged particles, and slowly rotated in order to optimize their measurements. Another new feature of the spacecraft:

Galileo carried a descent probe designed to drop into Jupiter's turbulent atmosphere, called the Jovian Atmospheric Probe at the bottom of the figure below.

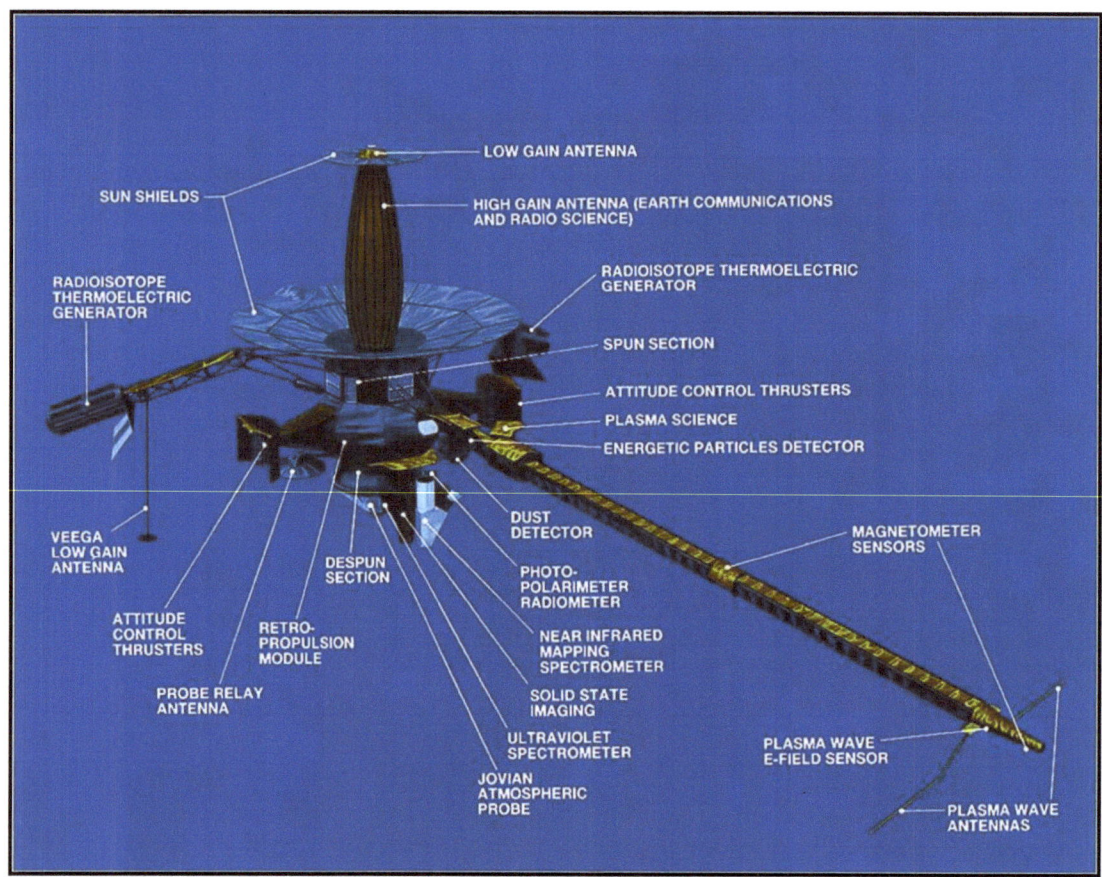

John Casani was the project manager from 1978, followed by Richard Spehalski (1988 to 1990, including the Galileo launch), Bill O'Neil, Bob Mitchell, Jim Erikson, Eilene Theilig, and finally Claudia Alexander. The spacecraft was built at JPL. Ames Research Center built the descent probe that went into Jupiter's atmosphere. Instruments were contributed by the Max-Planck-Institut (Germany); the Johns Hopkins University's Applied Physics Laboratory; the University of Colorado; JPL; the University of California Los Angeles; the Goddard Institute for Space Studies; the University of Iowa; and the National Optical Astronomy Observatories.

The image to the right shows the assembly of Galileo in JPL's Spacecraft Assembly Facility.

Disaster

Galileo was delayed by the Challenger explosion, and the mission was subsequently modified to accommodate new constraints on the Space Shuttle. One modification was to redesign the trajectory to fly past Venus to get an extra gravity assist, but at the cost of experiencing a relatively hot environment, as well as the extremely cold environment at Jupiter. The total launch delay was nearly 3 years.

The high gain antenna that had passed performance tests on the ground failed to open in flight, nearly two years into the mission. The antenna is shown below, in the Spacecraft Assembly Facility.

Here are excerpts from internal communications on this event:

"On April 11 [1991], Galileo's High-Gain Antenna (HGA) should have deployed. However, signals received from the spacecraft indicated that the antenna only partially deployed..."

"By April 30, the Anomaly Team initially speculated that "one or more ribs are probably restrained in the stow position, resulting in an asymmetrical partial deployment." At that time, this was the most probable scenario, but the cause was unclear."

Attempts to correct this situation continued through August, 1991, but to no avail. The rate that data could be returned would be reduced by nearly two orders of magnitude. The on-board data storage system was a digital tape recorder, with limited capacity.

The mission responded to this disaster by forming a tiger team that reprogrammed the on-board computers to take maximum advantage of data compression. It was hoped that the combination of optimized data transmission from the low gain system, plus on-board storage, would greatly improve the telecommunications performance of the mission.

To Make Matters Worse

It is a little-known fact that the tape system began to show signs of wear several years into the mission. One end of the tape was slipping, and may have become completely detached from the spool that held it. It was decided to refrain from completely rewinding the tape, in order to reduce the likelihood that the tape system would fail completely. It was understood on the project that the digital tape memory system could fail completely at any time.

How it Turned Out

Somehow, the tape system held together for the duration of the mission (14 years). Because of the extended mission duration, and software improvements onboard Galileo, an estimated 92% of the science data that was originally expected was actually returned from this mission. The tiger team and those who continued to perform mission operations had pulled off an extraordinary engineering accomplishment.

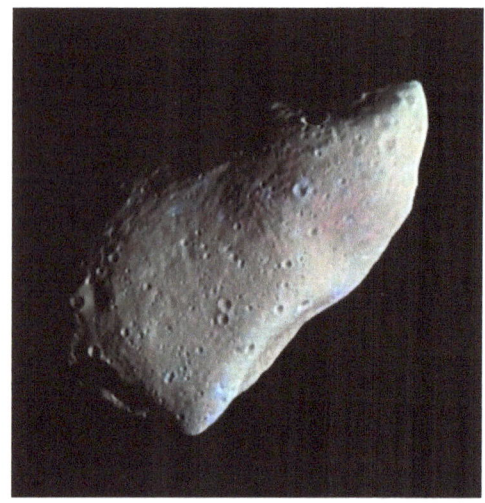

En route to Jupiter, Galileo flew close to two asteroids, the first such visits by any spacecraft. It encountered the asteroid Gaspra on October 29, 1991 and the asteroid Ida on August 28, 1993. Gaspra is an irregular body with dimensions about 12 x 7.5 x 7 miles. The picture to the left combines the highest-resolution image and color information obtained by Galileo, on October 29, 1991.

During the latter part of its cruise to Jupiter, Galileo observed the collisions of fragments of the comet Shoemaker-Levy with Jupiter. This is the first collision of two solar system bodies ever to be observed. Comet Shoemaker-Levy 9 consisted of at least 21 discernable fragments with diameters estimated at up to 2 kilometers, which collided with Jupiter from July 16 through July 22, 1994.

Galileo arrived at Jupiter on December 7, 1995, entering orbit and dropping its instrumented probe into the giant planet's atmosphere. The spacecraft then made about two and a half dozen orbits of Jupiter, usually flying close to one of its four major moons during each loop around the planet.

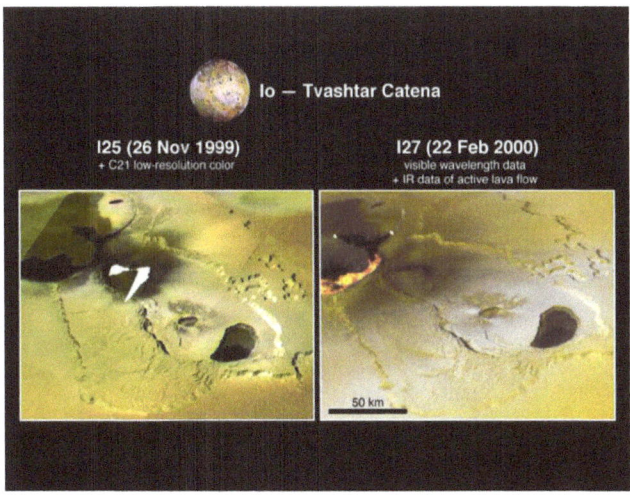

Scientists were surprised to find hot eruption sites scattered across Jupiter's moon Io. The detection of very high temperature volcanism-- hotter than any terrestrial lavas currently erupting--is one of the most spectacular discoveries by the Galileo mission. The images are of an area called Tvashtar and show lava fountains erupting on the surface. They were taken by the Galileo spacecraft on November 26, 1999. The active region is approximately 15 miles long and .5 miles high. The highest levels of radiant energy emitted from the lava was so intense that it overpowered the camera's detector and it is registered only in white.

Galileo discovered strong evidence that Jupiter's moon Europa (shown below) has a melted saltwater ocean under an ice layer on its surface. The spacecraft also found indications that two other moons, Ganymede and Callisto, have layers of liquid saltwater as well.

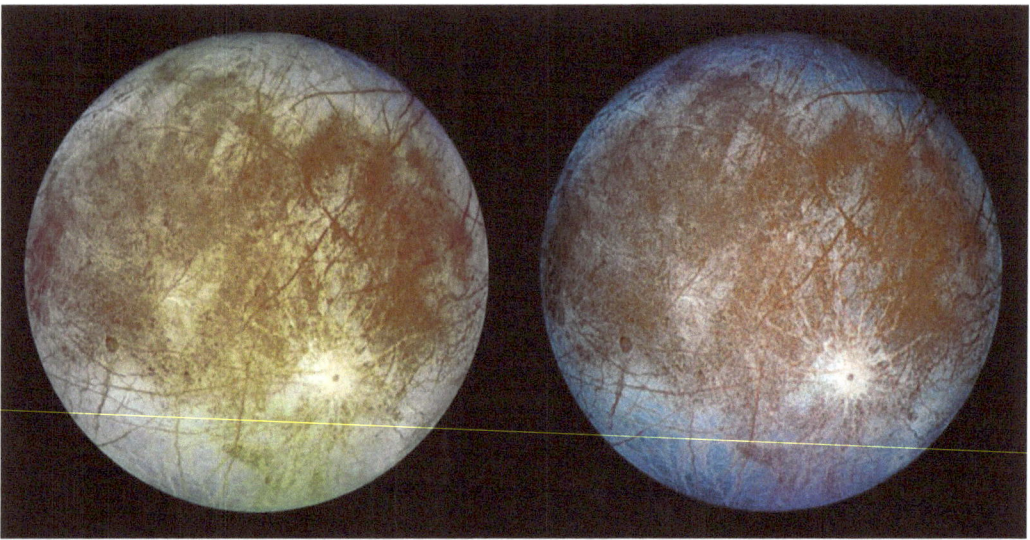

Other major science results from the mission include measurements of conditions within Jupiter's atmosphere, and discovery of a magnetic field generated by Ganymede.

The mission ended on Sept. 21, 2003, when the spacecraft plunged into Jupiter's atmosphere. This planned maneuver prevented the risk of Galileo drifting to an unwanted impact with the moon Europa, which may harbor a subsurface ocean.

Ulysses – Exploring the Sun

Ulysses was a joint mission between NASA and the European Space Agency (ESA) to study the Sun.

The spacecraft to the right was in the European assembly facility, before it was delivered to Kennedy Space Center.

From its inception in 1978 until 1984, it was known as the International Solar Polar Mission (ISPM), then an Italian astronomer suggested the name Ulysses.

Ulysses was launched on October 6, 1990 aboard Space Shuttle Discovery and sent towards Jupiter. Since no rocket engines are powerful enough to boost a spacecraft out of the Ecliptic Plane (where most planets and spacecraft orbit the Sun), Ulysses used Jupiter's gravity to bend its flight path downward and guide it onto the correct trajectory.

Ulysses was launched on the space shuttle Discovery (STS-41) on October 6, 1990 and deployed from the shuttle payload bay 160 nautical miles above Earth. The image to the left shows Ulysses leaving the Shuttle.

The purpose of Ulysses was to orbit the sun, passing over the north and south poles. It used nine instruments to study magnetic fields, solar wind, solar flares, as well as cosmic rays and cosmic dust. It takes about 6 years to complete each orbit around the Sun.

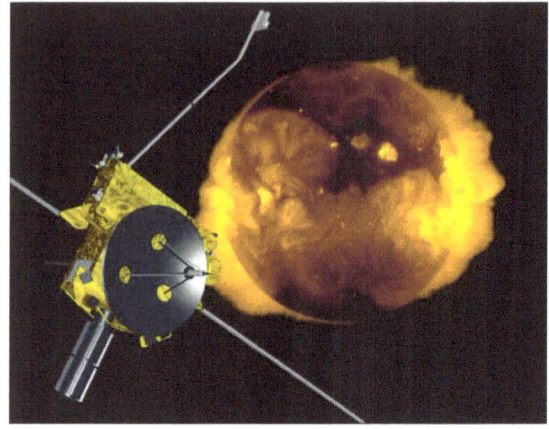

Ulysses was designed to complete a five-year mission, but NASA and ESA granted three extensions to the project, giving it time to complete two extra orbits of the sun. The picture to the left is an artist's conception of Ulysses, orbiting the Sun.

Ulysses flew through the tail of comet Hyakutake in May 1996. "The solar wind seemed to almost disappear and was replaced by gases not normally found in the solar wind, and the magnetic field in the solar wind was distorted."

At the time of the unexpected encounter, Ulysses was more than 300 million miles from Hyakutake (three times the distance from Earth to the sun), making it the longest comet tail ever recorded.

The power system on-board can no longer operate most instruments, so the project was officially terminated in 2008. The spacecraft is currently near Jupiter, on its way back to the Sun for another south polar pass.

Significance
Ulysses is the only mission that has ever achieved a polar orbit around the sun, yielding unique data on the primary object in our solar system.

Willis Meeks was the Project Manager, the first African-American to be a project manager at JPL.

Microwave Limb Sounder – Measuring Earth's Atmosphere

The Microwave Limb Sounder (MLS) experiments measure naturally-occurring microwave thermal emission from the limb (edge) of Earth's atmosphere to remotely sense vertical profiles of atmospheric gases, temperature, pressure, and cloud ice. The overall objective of these experiments is to provide information that will help improve our understanding of Earth's atmosphere and global change.

The first MLS experiment in space (UARS MLS) was on NASA's Upper Atmosphere Research Satellite (UARS) which launched 12 September, 1991.

The major objective of UARS MLS was to investigate the chlorofluorocarbon threat to the ozone layer, by providing global information on chlorine monoxide (ClO), the dominant form of chlorine that destroys ozone (O_3).

The image to the right was taken by UARS MLS, showing both ClO and the ozone hole.

UARS MLS generally provided daily measurements from 29 September 1991 through 15 March 1994 (although stratospheric water vapor measurements ceased on 15 April 1993).

After 15 March 1994 the measurements became increasingly sparse in order to conserve lifetime of the MLS antenna scan mechanism and UARS power. The last data were obtained on 25 August 2001.

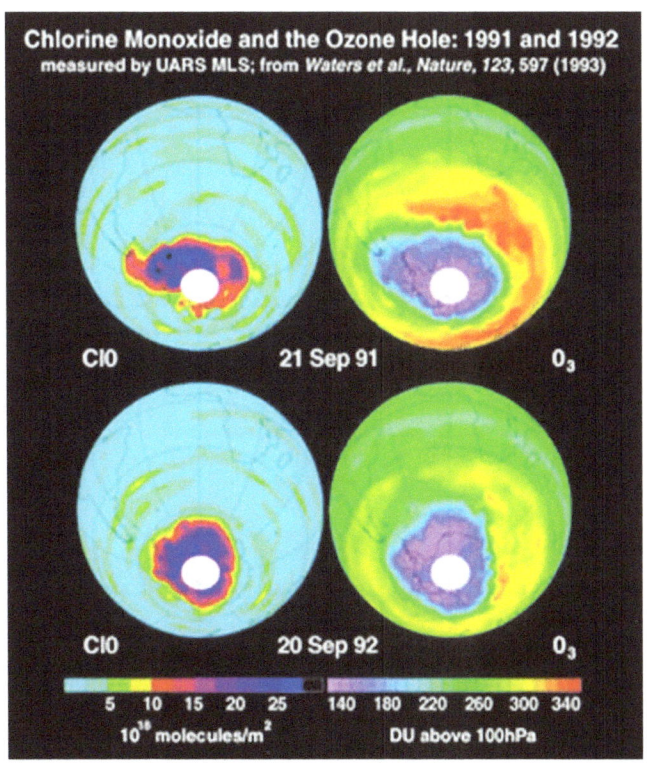

Results
Scientific understanding of the Ozone hole improved greatly. Public policy decisions were subsequently made to limit the use of chlorofluorocarbons to protect our environment and save human lives.

Magellan – Mapping Venus

Magellan was named after the sixteenth-century Portuguese explorer whose expedition first circumnavigated the Earth. Charles Elachi, later to become the JPL's director, did the initial studies for this mission in the early 1970s, but it took many years to get the mission approved.

The photo to the left shows Magellan in the Vertical Processing Facility at NASA's Kennedy Space Center on March 17, 1989, when it was being prepared for launch. Built partially with spare parts from other missions, the Magellan spacecraft was 15.4 feet long, topped with a 12 foot high-gain antenna. The spacecraft weighed a total of 7,612 pounds (including fuel) at launch.

The high-gain antenna, used for both communication and radar imaging, was a spare from the Voyager mission, as were Magellan's 10-sided main structure and a set of thrusters. The command data computer system, attitude control computer and power distribution units are spares from the Galileo mission to Jupiter. Magellan's medium-gain antenna is from the NASA/JPL Mariner 9 project.

First John Gerpheide and then Tony Spear were the Magellan project managers through launch. Dr. R. Stephen Saunders was the Magellan project scientist. Martin Marietta was the prime contractor for the Magellan spacecraft, while Hughes Aircraft was the prime contractor for the radar system. The spacecraft was launched aboard the Space Shuttle Atlantis on May 4, 1989 and it was then released from the cargo bay. The image to the right shows it being released.

Magellan arrived at Venus on August 10, 1990. By September 14, 1992 mapping coverage was 98% of the planet, with a resolution of approximately 100 meter. This is much better resolution than any other mission to Venus.

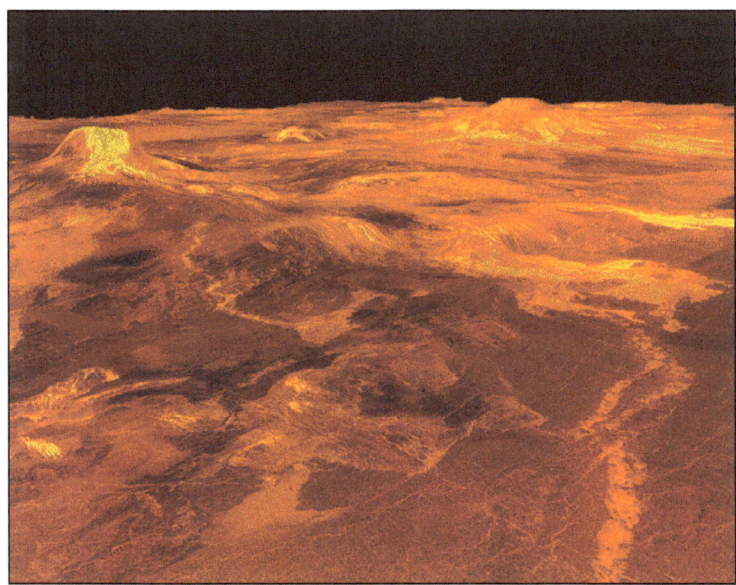

Here is a three- dimensional perspective view of the surface of Venus. Lava flows extend for hundreds of kilometers across the fractured plains shown in the foreground, to the base of Gula Mons.

Magellan synthetic aperture radar data is combined with radar altimetry to develop a three-dimensional map of the surface.

Discoveries

Scientists were intrigued by the distribution of volcanoes around Venus. On Earth, volcanoes occur in groups such as the so-called "Ring of Fire" around the Pacific Rim.

Venus, by contrast, is peppered with hundreds of thousands to millions of volcanoes distributed more or less randomly around the planet.

The image to the right shows three "pancake" domes, volcanic extrusions 39 miles wide, and half a mile tall.

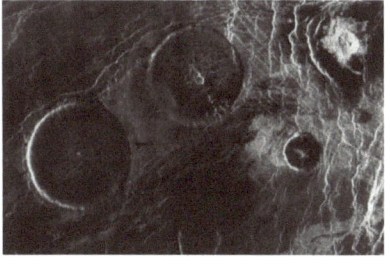

Scientists were also surprised to see huge channels thousands of kilometers long on Venus.

The image of the highland of Ovda Regio on Venus is shown below. It is made up of radar images collected during all three of Magellan's eight-month mapping cycles. Vertical exaggeration is 22.5 times. Ovda Regio covers an area 3,900 miles by 1,300 miles, rises over 10,000 feet above the surrounding plains, and is made up of complex ridge terrain, also know as tessera. At the bottom right of this scene, along the northern edge of the highland, are a series east-west trending ridges. In several places the ridges are flooded by smooth lava flows, indicating that volcanic activity postdates the formation of the ridges. The terrain in the interior of Ovda forms a chaotic pattern indicating multiple directions of deformation.

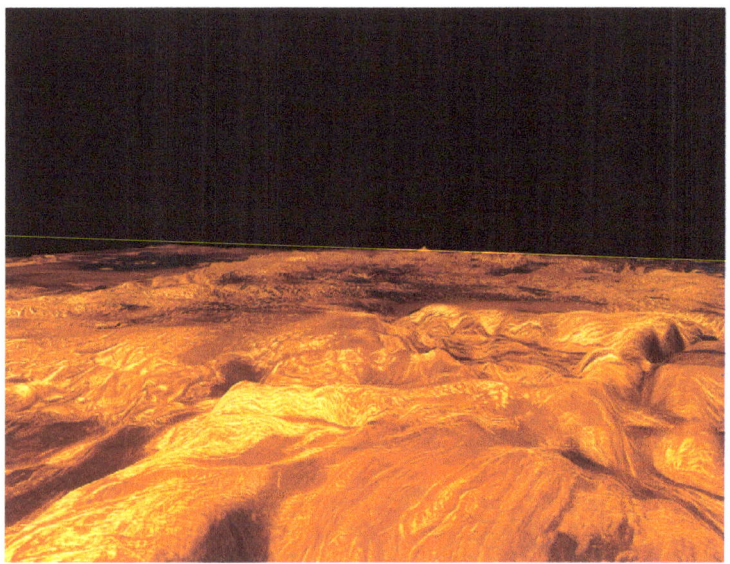

Magellan's data convinced scientists that Venus has a relatively young planetary surface, perhaps about 500 million years old. Since Venus formed at the same time as Earth 4.6 billion years ago, some event or events 500 million years ago must have resurfaced the planet. Magellan's maps show no telltale signs of past major water bodies such as shorelines or ocean basins. Also, surface features show no evidence of being eroded by water -- although there is evidence of wind erosion in the form of numerous sand dunes and wind streaks.

Wide Field and Planetary Camera – The Sharp Eye of Hubble

JPL's first Wide Field Planetary Camera launched with NASA's Hubble Space Telescope on April 24, 1990. Hubble is shown to the right.
(**Image credit**: NASA and STScI)

Hubble contains a 2.5-meter-diameter (roughly 8-foot) mirror that collects light from extremely distant objects in deep space. This light is brought to focus at a particular point where any one of five onboard instruments can turn it into pictures that are sent by radio to Earth. For years the main instrument used for taking general pictures of stars, galaxies and planets has been JPL's Wide Field and Planetary Camera.

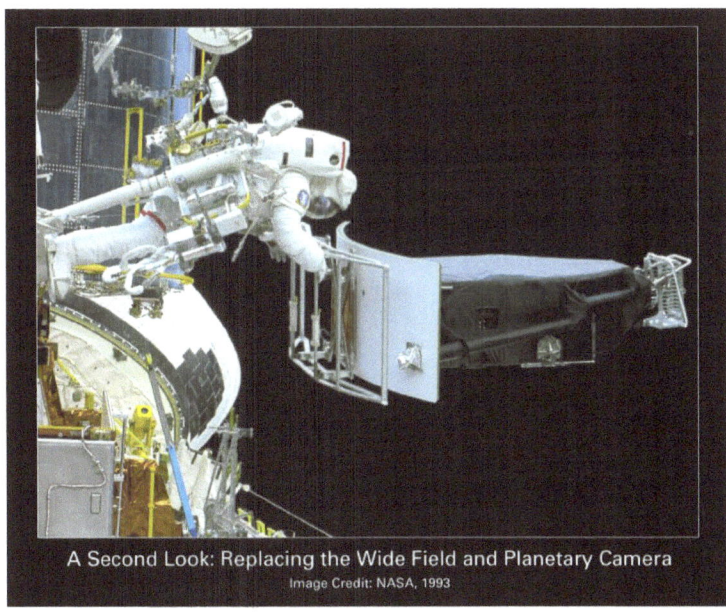

A Second Look: Replacing the Wide Field and Planetary Camera
Image Credit: NASA, 1993

An optical flaw in Hubble's main mirror compromised the entire project. JPL then produced its Wide Field and Planetary Camera 2 with special optics to correct the space telescope's vision. This replacement camera was launched on December 2, 1993, and was installed on the Hubble telescope by spacewalking astronauts (shown below). This brought Hubble's vision to near-perfect focus. Hubble went from being a failed mission to one of the greatest missions of its kind in NASA's history.

Image credit: NASA and STScI.

The Wide Field Planetary Camera has provided many of the iconic images from the Hubble mission.

Below is the "Pillars of Creation" image of the Eagle Nebula. It shows columns of cool interstellar hydrogen gas and dust, which are believed to be incubators for new stars. The Eagle Nebula is 6,500 light-years away in the constellation Serpens. This image was taken on April 1, 1995 by WFPC 2. **Image credit**: NASA and STScI

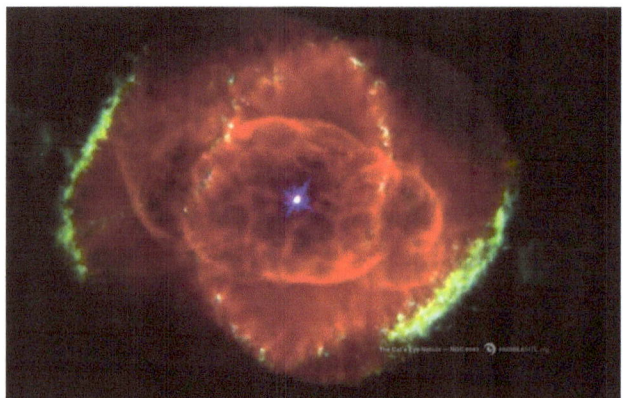

The "Cat's Eye Nebula," is a complex structure of concentric gas shells, jets of high-speed gas and unusual shock-induced knots of gas which is 1,000 light-years from Earth.

It is thought to be the product of two stars orbiting each other, in such close proximity that they appear as one one star in this image.

Image credit: NASA and STScI.

his picture of the Crab Nebula shows the six-light-year-wide expanding remnant of a star's supernova explosion. Japanese and Chinese astronomers recorded this violent event nearly 1,000 years ago in 1054. The orange filaments are the tattered remains of the star and consist mostly of hydrogen.

Image credit: NASA and STScI.

Larry Simmons managed this project. Wide Field and Planetary Camera 2 returned to Earth aboard Space Shuttle Atlantis in May, 2009, and will go to the Smithsonian Air and Space Museum.

More on JPL Management

In 1991, Lew Allen retired and Edward C. Stone, the Voyager project scientist, became JPL's director.

The 1990s were a period when many new management philosophies were promoted throughout NASA. Under Ed Stone, JPL went through:

- Total Quality Management
- Process Reengineering
- Faster, Better, Cheaper
- The emergence of extensive new programmatic directives from NASA
- The formal documentation of JPL's management practices for flight projects
- The formal documentation of JPL's engineering principles for flight projects.

Ed Stone also made conscientious efforts to be accessible to all levels of employee. He would schedule lunch meetings with ten or so employees, selected from various JPL organizations. I remember attending one such lunch, along with a mix of new hires and veteran employees from several different Sections of the Laboratory. Ed was excellent at asking questions and listening to people.

Topex/Poseidon – Investigating the Oceans

Topex/Poseidon revolutionized our understanding of the oceans. Planning for this mission began in 1983, with actual launch occurring on August 10, 1992.

Topex/Poseidon, shown to the right, was a joint U.S./French mission. Charlie Yamerone managed the JPL part of this project.

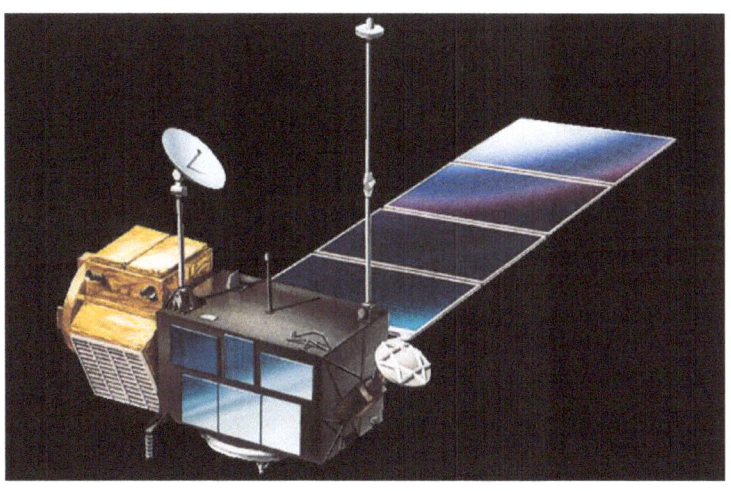

Topex provided ocean scientists with a unique, revolutionary view of our ocean waters which are stirred and mixed by mighty currents, distributing heat across the globe and regulating our climate.

Major achievements of the mission include:

- Continuous global coverage of the ocean surface topography, leading to a new understanding of ocean circulation that affects climate change.

- Provided an effective way to monitor El Nino and La Nina (see image to left).

- Improved determination of deep ocean tides and their effects on ocean general circulation.

Topex/Poseidon monitored changes of the sea level, one indicator of the of global temperature change.

Topex/Poseidon also monitored the temperature on the surface of the ocean. This helped scientists who study hurricanes. The image to the right shows the hurricane Wilma, in the Gulf of Mexico.

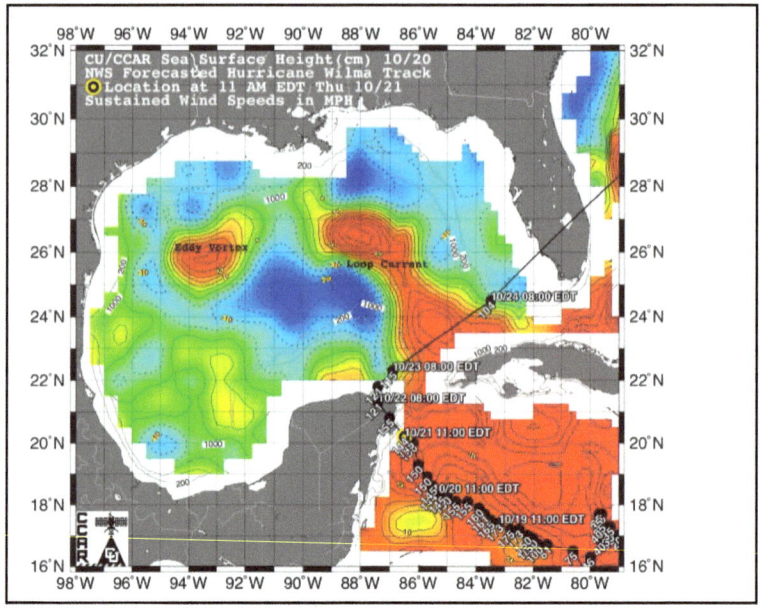

Topex/Poseidon greatly exceeded its design life. Launched in August, 1992, it began full operations in September of that year, and was expected to last for three years. It was decommissioned in December, 2005, more than ten years beyond its design life.

Mars Observer – Disaster at Mars

The NASA budget for planetary science in the 1980s was insufficient to support additional "flagship" missions such as Galileo. NASA officials decided upon a strategy that was somewhat similar to the Mariner program, which would produce a series of low-cost planetary science missions to be called the Planetary Observer line. Cost savings would come from (1) using the Space Shuttle as a low cost launch vehicle; (2) using commercially-built Earth orbiter class spacecraft; and (3) selecting only scientific instruments that were already mature. The Observer line would not develop new technology.

Mars Observer was the first of this series of missions. Unfortunately, this strategy was violated early and often through the life cycle of this mission, resulting in the highest cost overrun in JPL's history (in percentage terms), and a mission failure upon arrival at Mars.

Mars Observer was approved in 1985 for a 1990 launch on the Space Shuttle. The spacecraft was based on a commercial Earth-orbiting communications satellite that would be converted into an orbiter for Mars. The image to the right is the spacecraft, being assembled at the contractor site.

The payload of science instruments was designed to study the geology, geophysics and climate of Mars.

This is where the program strategy started to be violated. The payload was extensive and sophisticated, including exciting new instrument performance capabilities. "Mars Observer is the most complex mission we have ever flown to Mars", according to Dr. Wesley T. Huntress, Director of NASA's Solar System Exploration Division.

The loss of Challenger led to a lengthy delay, followed by a decision to launch Mars Observer on a Titan III expendable launch vehicle in September, 1992. The switch in launch vehicles meant an immediate increase in the budgeted cost for launch services. Also, a different launch environment for the spacecraft and instruments meant there had to be changes in design, and particularly changes in the verification and validation effort. Finally, the two year launch delay added thousands of work years to the cost of this project. Originally budgeted at $212 million, Mars Observer wound up costing $813 million.

But it did launch, and seemed to perform well on the flight to Mars.

Mars Observer disappeared Aug. 22, 1993, two days before entering Mars orbit. The flight team had just ordered the spacecraft to pressurize its propulsion system. An independent failure investigation board concluded that a tiny amount of leakage through some of the valves had most likely caused a small explosion, throwing the spacecraft permanently out of control. This determination was somewhat speculative, however, as JPL had received no telemetry at all from the spacecraft.

Failure review efforts were quite thorough, with lessons learned being recorded in the NASA "Lessons Learned Information System" as late as 1997. Among many other findings, this observation was made:

> "too much reliance was placed on the heritage of spacecraft hardware, software and procedures for near-Earth missions, which were fundamentally different from the interplanetary Mars Observer mission"

Regardless of cause, Mars Observer's loss was traumatic for people at JPL. It was JPL's first complete mission failure since 1967.

Mars Observer was also the first Mars mission in 17 years, and many JPLers are Mars enthusiasts. Recovering from the Mars Observer failure became a major priority for both NASA and JPL.

Bill Purdy and then David Evans were the Project Managers through launch. Glen Cunningham was the Project Manager during operations. RCA was selected to build the spacecraft. RCA was subsequently acquired by Lockheed Martin.

A Back Story
A few months after the loss of Mars Observer, I presented a paper on "Risk Management" to a large audience of managers at JPL. Glenn Cunningham, who was a manager of Mars Observer and subsequently would be the manager of the Mars Global Surveyor, endorsed my recommendation of a formal risk management process on flight projects. He piloted this approach on his next project, and JPL has had formal risk management plans and procedures on all flight projects since that time.

NASA Scatterometer – Measuring the Wind

The NASA Scatterometer (NSCAT) instrument was designed and built by JPL and flown on Japan's Midori satellite (previously known as the Advanced Earth Observation Satellite (ADEOS)), the largest satellite ever developed by that country.

The NSCAT instrument has six 3-meter long, stick-like antennas which collect data with a resolution of 50 km.

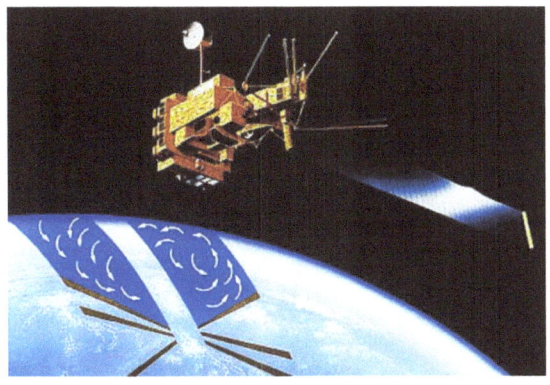

Following launch from Japan's Tanegashima Space Center on August 17, 1996 (the evening of August 16 in U.S. Pacific Daylight Time or Eastern Daylight Time), the satellite circled Earth in an orbit that took it close to the planet's north and south poles.

Every two days, under all weather and cloud conditions, NSCAT measured wind speeds and directions over at least 90% of the Earth's ice-free oceans. Since oceans cover approximately 70% of Earth's surface, NSCAT played a key role in scientists' efforts to understand and predict complex global weather patterns and climate systems.

NSCAT used eight antenna beams to scan two wide bands of ocean, one on each side of the instrument's orbital path. NSCAT transmitted short pulses of microwave energy to probe ocean surfaces and then measured the reflected or backscattered power. Variations in the magnitude of this backscattered power are caused by changes in small (centimeter-sized), wind-driven waves. Using a method called *Doppler processing* (a change in the observed frequency of the radio waves due to relative motion of source and observer), the measured backscattered power was separated into cells at specific locations on Earth's surface; these were then transmitted to the ground for processing. During ground processing, wind direction and speed was determined from these variations. Within two weeks of receiving the raw data, the ground system processed wind measurements.

The NASA Scatterometer Project was managed by Firouz Naderi.

NSCAT yielded 268,000 measurements of ocean winds each day, covering more than 90 percent of Earth's ice-free seas. The image below shows an NSCAT product.

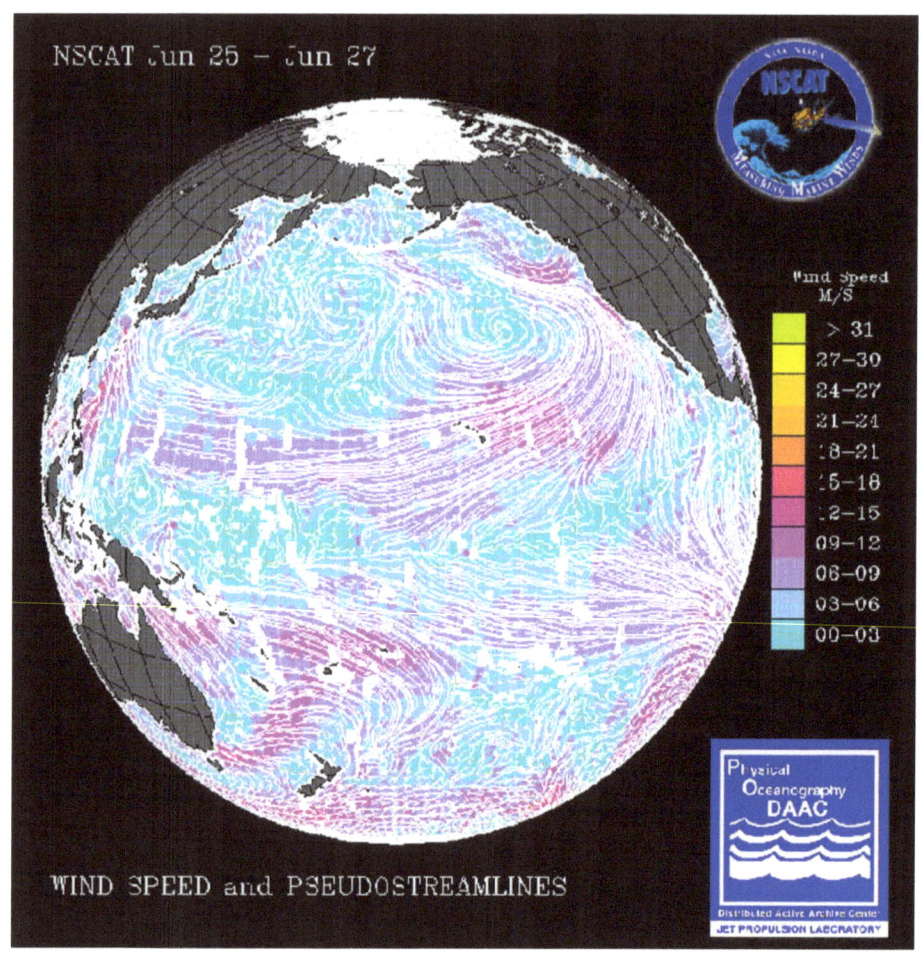

The Midori satellite lost power in July, 1997, ending NSCAT's mission.

Results

The value of Scatterometer data in measuring wind and predicting weather was established by NSCAT. Consequently, NASA approved a rapid replacement mission called the Quick Scatterometer, or QuikScat, to take the place of NSCAT.

Mars Global Surveyor – Discoveries at Mars

Recovering from the loss of Mars Observer was a top priority.

A complete set of spares for a re-flight of the mission existed, but the cost of a second Titan III launch rocket was too expensive. NASA Director Dan Goldin ordered the use of the much less expensive (and much smaller) Delta 2. Lockheed-Martin Astronautics of Denver proposed re-flying five of Mars Observer's eight experiments using the spare electronics, a new spacecraft bus and the aerobraking technique. Instruments came from Goddard (laser altimeter and a magnetometer); Malin Space Science Systems (camera); Arizona State University (spectrometer); and Stanford (radio science). This became the Mars Global Surveyor mission.

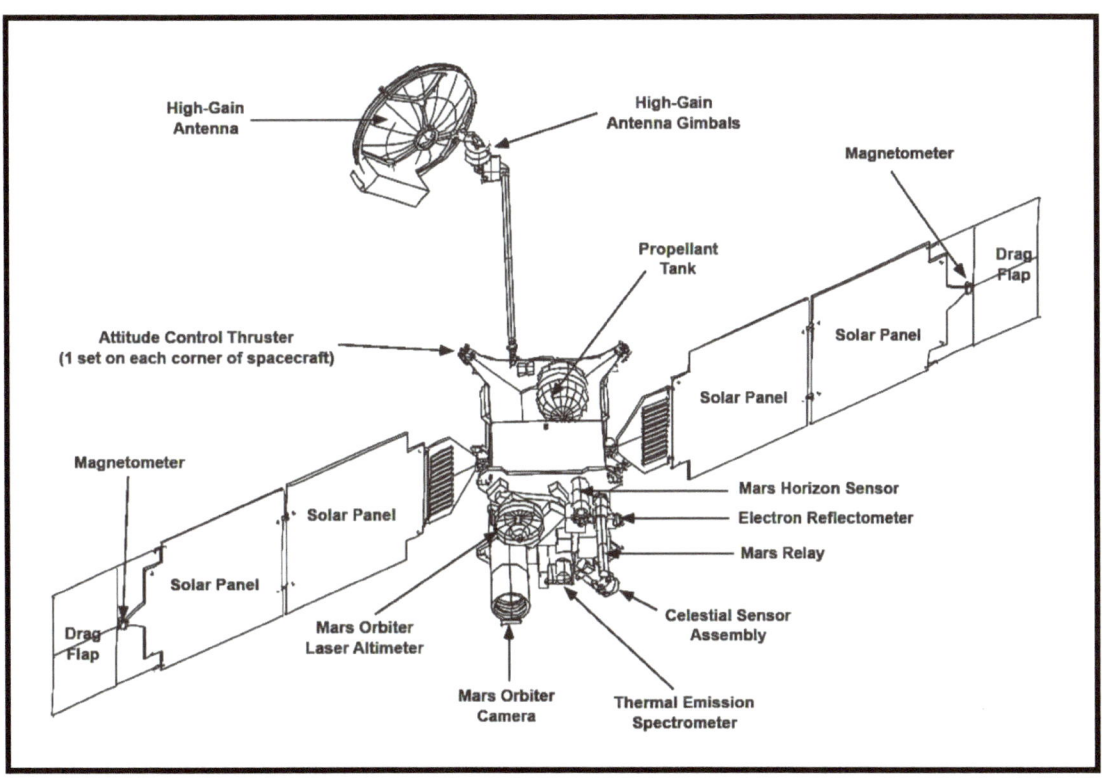

It should be noted that by re-flying previously built and tested instruments, the MGS project came close to fulfilling the intentions of the Observer Program that officially died with Mars Observer.

Mars Global Surveyor was launched November 7, 1996, with a cost under-run of about $2 million. It started aerobraking in September 1997, and its flight team expected to achieve its science orbit six months later.

But during one of the aerobraking passes, one of the solar panels started flexing in the air flow too much, and the team had to slow the process to protect it. The flaw was traced to a part that had cracked shortly after launch.

The picture to the right is an artist's concept of MGS while orbiting Mars.

Glenn Cunningham, the MGS Project Manager, was faced with a decision on how aggressively should the spacecraft aerobrake, if at all, under these adverse circumstances. His decision was to continue aerobraking, but at a more relaxed pace than originally planned. Consequently aerobraking continued until February 1999, nearly one year later than planned.

The MGS instruments worked perfectly throughout the mission, and the mission extended years beyond the four year prime mission of the spacecraft. Still operating until November 2, 2005, it produced mineralogical and magnetic field maps of the Mars surface. It also provided thousands of spectacular high-resolution images. Its data suggested Mars had once had plate tectonics as Earth still does. MGS returned more data on Mars than all other Mars missions combined, until Mars Reconnaissance Orbiter arrived at Mars a decade later.

One of the more spectacular discoveries was the apparent discovery of a stream of water on the surface of Mars. The images below show a comparison of a gully site as it appeared on December 22, 2001 (left), with a mosaic of two images acquired in August, 2005.

"These observations give the strongest evidence to date that water still flows occasionally on the surface of Mars," said Dr. Michael Meyer, lead scientist for NASA's Mars Exploration Program, Washington.

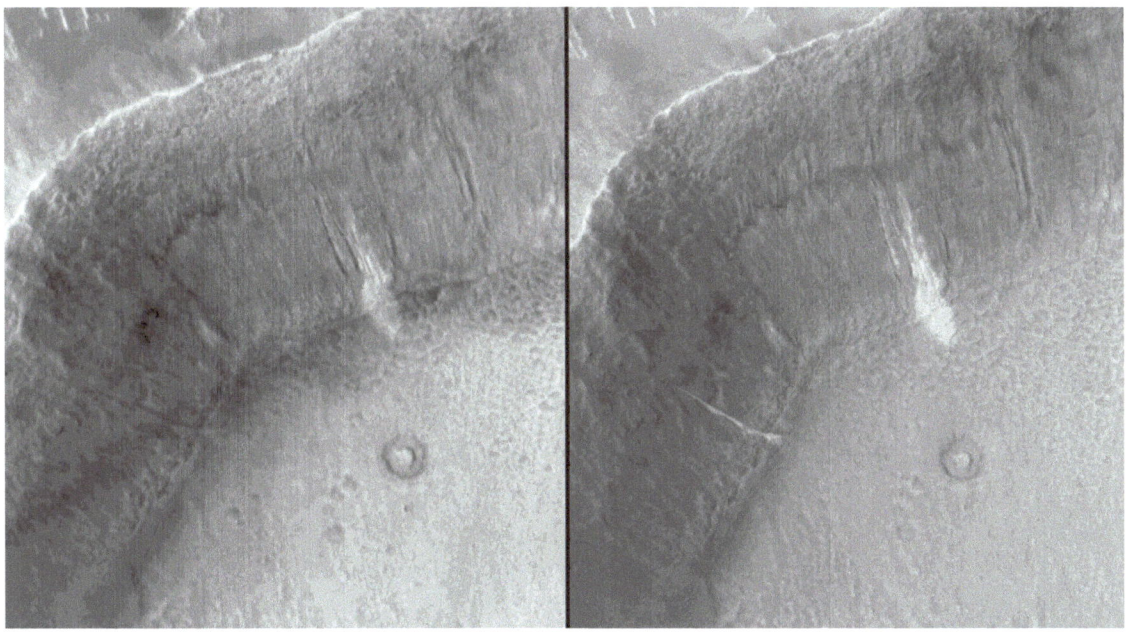

High resolution imagery supported Mars lander missions, such as the Mars Exploration Rover. The image to the right was used to fine tune the landing site for the Phoenix Lander.

MGS also saw duty as a telecommunications relay point for Mars Rovers.

Mars Global Surveyor last communicated with Earth on November 2, 2006.

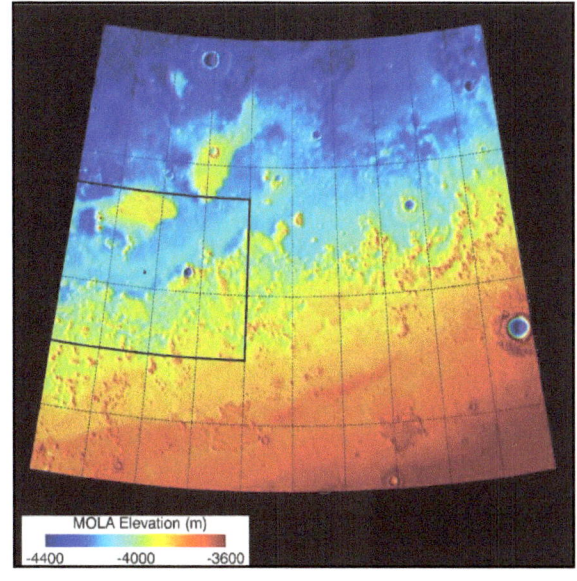

Mars Pathfinder – The Amazing First Rover

The Challenge

For a total cost of less than one 10th the budget of Viking, JPL was directed to build and land a small, short-lived spacecraft on Mars. Pathfinder would also deliver a small rover to the surface, which was supposed to operate for one week. Tony Spear was asked if he could lead such a mission, and he agreed, despite the absence of any of the usual preliminary studies and assessments.

The Reaction within JPL

Pathfinder was considered too risky to succeed by most people. Tony Spear and his team were much admired for their courage and originality, but not their good sense, because:

- It had been more than twenty years since the Viking lander, so there was no real inheritance in terms of parts, designs, or even personnel
- They would not have sufficient resources, even though in some respects Pathfinder would be more complex than Viking
- The spacecraft was going to crash land on airbags at nearly 80 miles per hour
- If those airbags landed on sharp rocks, or on a steep slope, the spacecraft could have been damaged, probably catastrophically
- The rover had never been used on another planet
- If anything went wrong with any of several systems, the entire mission would fail
- Public scrutiny would be intense. A previous Mars mission (Mars Observer) had been lost, and the NASA Director (Dan Goldin) was sharply critical of JPL's conduct of that mission.

Upping the Ante

The Pathfinder team appreciated the admiration for their courage and originality, but disagreed with the consensus about their good sense. They moved the entire team into a secure building (the second floor of the Space Flight Operations Center, which was restricted space, even to most JPL employees). The image to the right shows the lander in a test area.

Then they worked out a mission design so that the landing would take place on July 4, and would be shown as a live event on a giant screen at the Rose Bowl, where roughly 70,000 from the Pasadena area would gather to see a traditional July 4th fireworks display.

Test Failure

Initially, the Pathfinder lander would arrive on Mars using a small parachute and a set of airbags that were pretty similar to the airbags in your automobile. Prototypes were built and tested early in the project life cycle.

Unfortunately, the prototype failed during testing. This photo was taken in August 1993 at Sandia National Laboratory's Coyote Canyon aerial cable test facility in New Mexico. Cables were stretched between two small mountains to hoist and release this 3/8 scale airbag system prototype.

The airbags were then stiffened and a small retro-rocket was added to the entry-descent-landing system, at which point the airbags worked.

The parachute system was also tested, since the atmosphere on Mars is much less than on Earth. The parachute system also failed, and very careful engineering was required to modify terrestrial technology to work in a Mars environment.

The lander was formally named the Carl Sagan Memorial Station following its successful touchdown.

A diagram of the lander is shown to the right.

Upon landing, the lander weighed 816 pounds. With petals unfolded, the lander is 9 feet across. The mast fully extends to 5 feet.

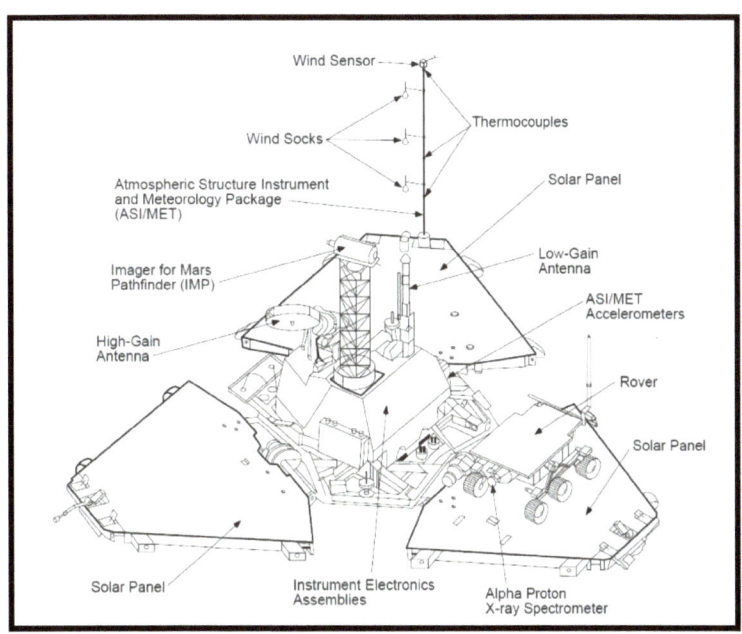

The rover was named Sojourner after American civil rights crusader Sojourner Truth. It weighed only 23 pounds. When deployed, it was 2 feet long, 1.5 feet wide, and 1 foot tall. A diagram of the rover is shown below.

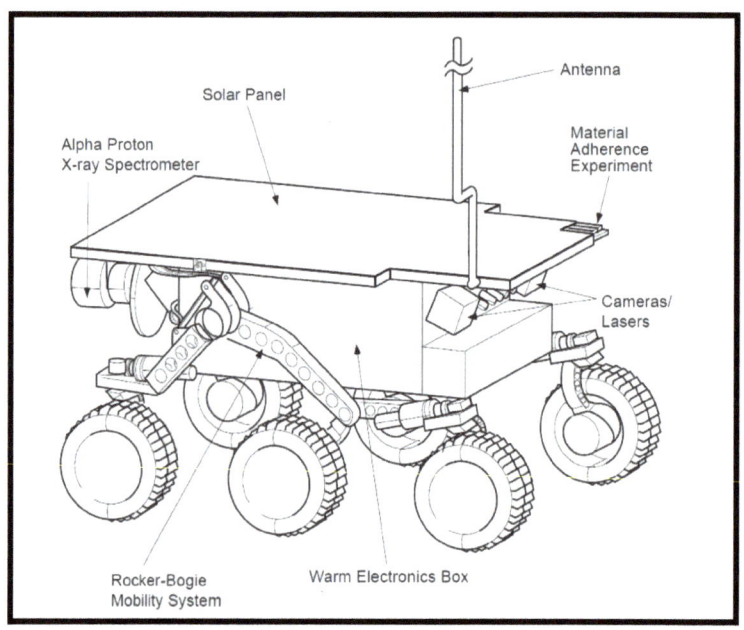

There was barely enough time to complete the design, fabrication, testing, redesign and retesting that was needed to take Pathfinder to the launch pad. But the mission was launched in December 1996 and landed on Mars on July 4, 1997.

The audience at the Rose Bowl was electrified by the sight of a live event on the surface of Mars, as were millions of other people around the world.

Prior to Pathfinder, popular NASA missions would get several million internet hits. Pathfinder got hundreds of millions of hits during its brief time on Mars.

The landing site, an ancient flood plain in Mars' northern hemisphere known as Ares Vallis, is among the rockiest parts of Mars. It was chosen because scientists believed it to be a relatively safe surface to land on and one which contained a wide variety of rocks deposited during a catastrophic flood. The image below shows a view of the Ares Vallis, with the rover in the foreground in its stowed configuration.

Here is an image of Sojourner as it leaves the lander.

Here is an image of the Carl Sagan Memorial Station, from the viewpoint of Sojourner.

From landing until the final data transmission on September 27, 1997, Mars Pathfinder returned 2.3 billion bits of information, including more than 16,500 images from the lander and 550 images from the rover, as well as more than 15 chemical analyses of rocks and soil and extensive data on winds and other weather factors.

Findings from the investigations carried out by scientific instruments on both the lander and the rover suggest that Mars was at one time in its past warm and wet, with water existing in its liquid state and a thicker atmosphere.

Notes on Management
Mars Pathfinder was completed within budget. This performance was due to: team co-location; excellent schedule management; cutting corners on formal documentation; use of parts that became available from other missions; strenuous effort by team members, who often logged 80 hour work weeks for months at a time. Some of these successful actions could be transferred to other projects; some would be unique to Pathfinder.

Personal Observation
Prior to Pathfinder, morale at JPL was low, due to Mars Observer and other setbacks. When Pathfinder succeeded, morale was greatly restored throughout the Lab.

Cassini-Huygens – Exploring Saturn

Cassini-Huygens is one of the most ambitious missions ever launched into space. Loaded with an array of powerful instruments and cameras, the spacecraft is capable of taking accurate measurements and detailed images in a variety of atmospheric conditions and light spectra.

The Cassini managers were John Casani, followed by Richard Spehalski (through launch), then Robert Mitchell (during operations).

The spacecraft has two elements: The Cassini orbiter and the Huygens probe.

The image to the right shows the Cassini spacecraft at JPL in October 1996. The gold circular object is a model of the Huygens probe. The entire craft is being loaded in the vacuum chamber of the environmental test laboratory. People are working on the lower part of the spacecraft.

It is equipped for 27 diverse science investigations, performed by its 12 science instruments. The science instruments include "remote sensing" devices (e.g., cameras, spectrometers, radar) and "direct sensing" devices (e.g., able to measure magnetic fields around the spacecraft, and detect particles).

Instruments were built by: the French Space Agency; JPL; Southwest Research Institute; the German Max Plank Institute; the Goddard Spaceflight Center; the University of Arizona;

Britain's Imperial College; Johns Hopkin's Applied Physics Laboratory; the University of Iowa; and the University of Colorado.

Cassini launched on October 15, 1998. In order to get to Saturn, Cassini picked up speed from gravity assists from Venus (twice, on April 26, 1998 and on June 24, 1999), and from Earth (August 18, 1999), and from Jupiter (December 30, 2000).

The image to the right shows Jupiter's magnetic fields, as they were mapped by Cassini during its flyby.

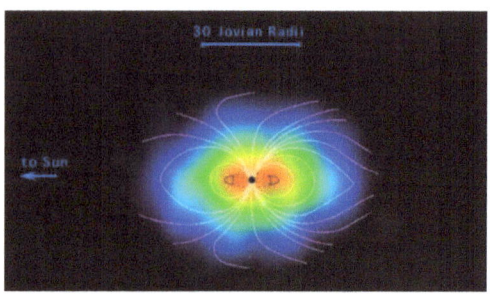

Saturn orbit insertion started on July 1, 2004. (see the image of Saturn, below.)

Huygens Probe

Huygens was released on December 24, 2004 and used a parachute to enter the murky atmosphere of Titan on January 14, 2005. The Huygens probe had six instruments, which provided our first images of the surface of the largest moon in our Solar system.

This is the Huygens probe's view of Titan from an altitude of 6 miles. The images that make up this view were taken on Jan. 14, 2005, with the descent imager/spectral radiometer onboard the European Agency's Huygens probe.
Image Credit: ESA/NASA/University of Arizona

Cassini completed its initial four-year mission to explore the Saturn System in June 2008.

The Cassini spacecraft is still healthy at the time of this writing (January, 2009), and is engaged in a follow-on phase, called the Cassini Equinox Mission.

The mission's extension, through September 2010, is named for the Saturnian equinox, which occurs in August 2009 when the sun will shine directly on the equator and then begin to illuminate the northern hemisphere and the rings' northern face. Cassini will observe seasonal changes brought by the changing sun angle on Saturn, the rings and moons, which were illuminated from the south during the mission's first four years.

Cassini is a very active mission, due to the complexities of navigating around Saturns many moons and rings. Orbital maneuvers occur almost weekly. In February, 2009 Cassini completed "orbital trim maneuver" number 181. Encounters with moons have occurred nearly every month since Cassini arrived. Many repeat visits are planned for Cassini at the moons Titan and Enceladus –important targets of the Equinox Mission.

Enceladus

Small, icy Enceladus is of great scientific interest. First, it is the brightest object in the solar system, reflecting nearly 100% of the light that strikes its surface. Even more fascinating it is surprisingly active.

Cassini discovered an icy plume shooting from this moon , and subsequent observations have revealed the spray contains complex organic chemicals. This false-color view was created by combining three clear filter images taken at nearly the same time on November 27, 2005 at a distance of approximately 92,000 miles.

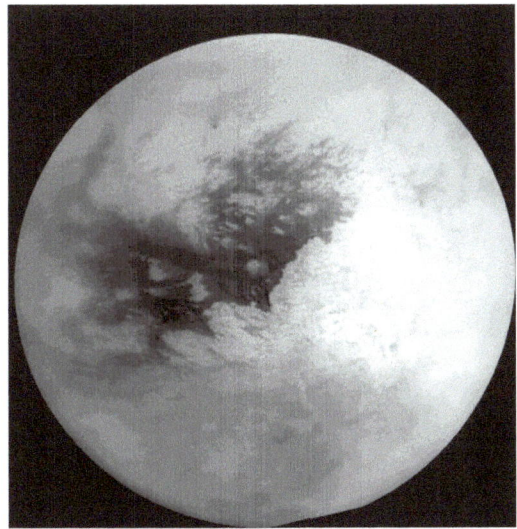

Tidal heating keeps it warm, and hotspots associated with the fountains have been pinpointed. With heat, organic chemicals and (potentially) liquid water, Enceladus could be a place where life might evolve. Questions surrounding Enceladus' "astrobiological potential" are at the heart of many investigations being conducted in the Equinox Mission.

Titan

Cassini catapulted our knowledge of giant, haze-enshrouded Titan into a whole new realm. Although it is classified as a moon, Titan's diameter is 3,200 miles right between the diameters of Mercury (2,400 miles) and Mars (4,200 miles).

Cassini investigates the structure and complex chemistry of Titan's thick, smog-filled atmosphere. On the frigid, alien surface, the spacecraft and its Huygens probe revealed vast methane lakes and widespread stretches of wind-driven hydrocarbon sand dunes. Cassini researchers also deduced the presence of an internal, liquid water-ammonia ocean.

Titan's surface is changing. There are signs of seasonal climate changes such as storms, flooding, and changes in lake levels, as well as evidence of volcanic activity. These mosaics of the south pole of Saturn's moon Titan, made from images taken almost one year apart, show changes in dark areas that may be lakes filled by seasonal rains of liquid hydrocarbons. The images on the left were acquired July 3, 2004. Those on the right were taken June 6, 2005. The 2005 images show new dark areas, which are surface liquids (see circles). The very bright features are clouds, which change rapidly on timescales of hours and appear in different places from day to day.

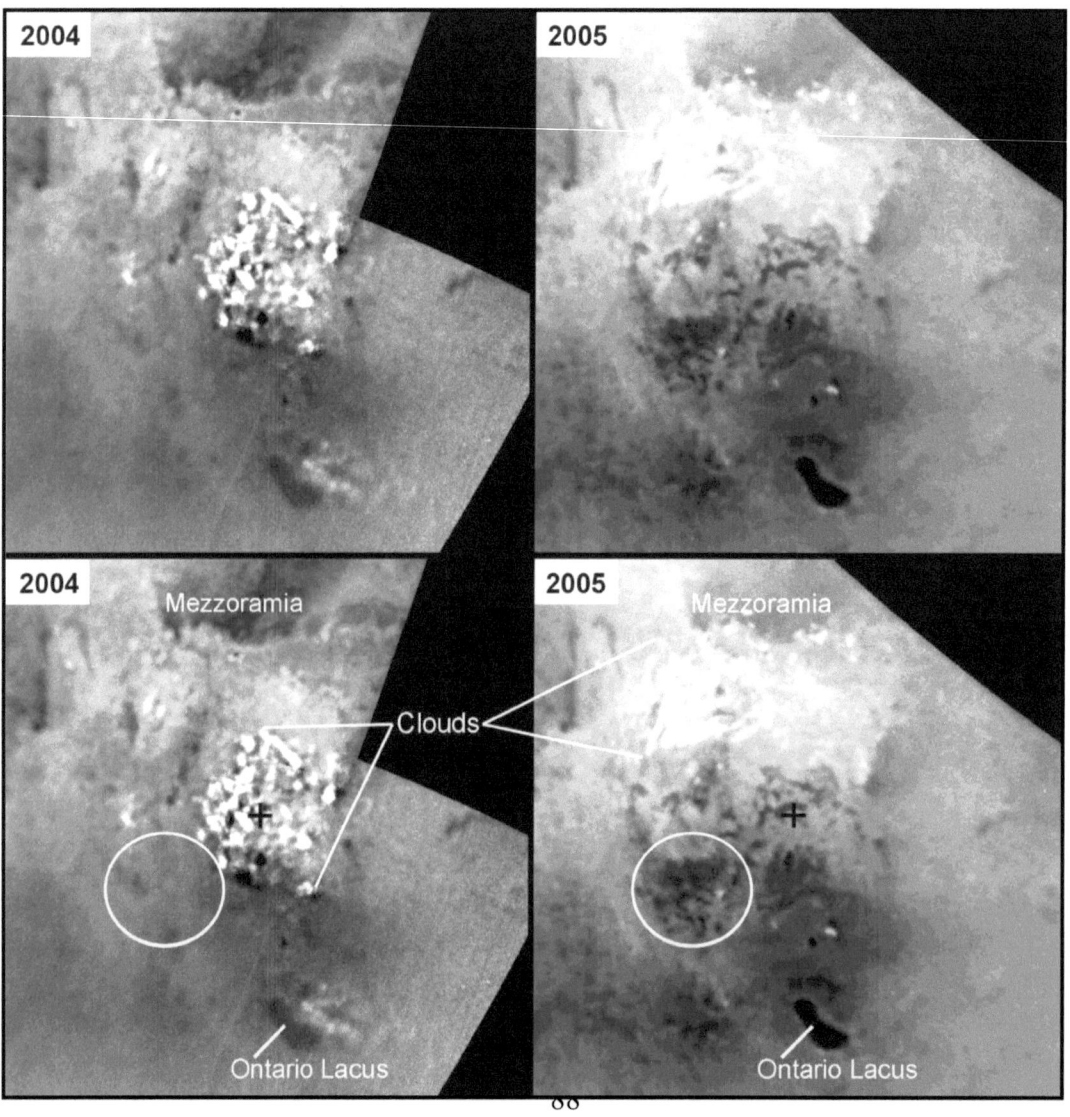

Deep Space One – Testing New Technologies

A NASA program was created at JPL called New Millennium which was designed to flight-test new technologies for future space and Earth-observing missions. The first flight project created under New Millennium was Deep Space 1, a spacecraft built to test a dozen new technologies including an ion engine.

Launched October 24, 1998, from Cape Canaveral, Florida, on a Delta II rocket, Deep Space 1 carried out most of its technology testing in the two months immediately after launch.

The ion engine was used to add thrust as the spacecraft orbited the Sun. The picture below shows the Deep Space 1 ion engine.

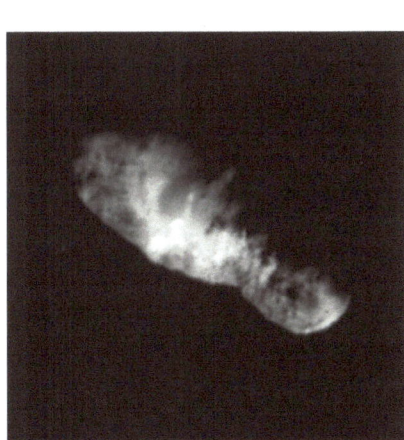

On July 29, 1999, the spacecraft flew by asteroid 9969 Braille. Since its primary mission ended in September 1999, Deep Space 1 went on an extended mission flying by comet Borrelly in September 2001 (shown to the right). The space probe's close encounter with comet Borrelly provides the best-resolution pictures of any comet to date. Deep Space 1 whizzes by just 1,400 miles from the rocky, icy nucleus of the 6-mile-long comet.

Dave Lehman was the Project Manager. Spectrum Astro Incorporated built the spacecraft, which was integrated and tested at JPL.

Deep Space 2 – Diving into Mars

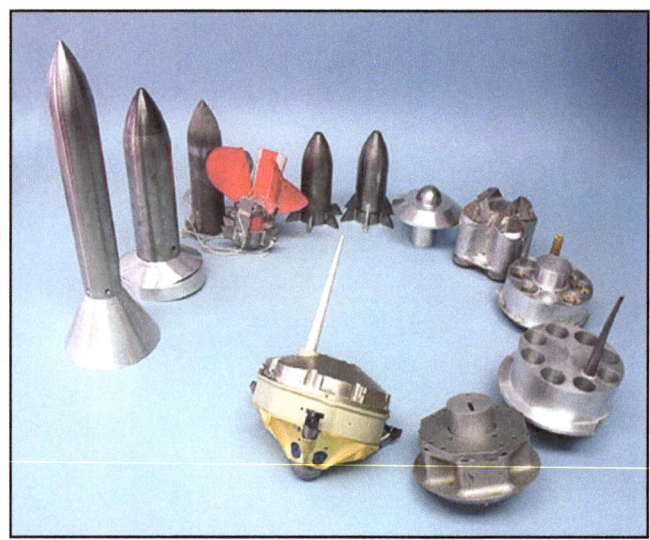

Deep Space 2 was a payload which was one of the more unlikely flight projects that JPL has attempted. The idea was to dive straight down from orbit at high speed and to plant a seismic sensor under the surface of Mars.

Four years of technology development produced a great many designs, as shown in the figure to the left.

Finally, two miniature probes, each weighing just 5.3 pounds, were launched aboard the Mars Polar Lander (see the next mission). An image of one of these probes is shown to the right.

Upon arrival just above the south polar region of Mars on December 3, 1999, the basketball-sized shells were released from the main spacecraft and plummeted through the atmosphere, hitting the planet's surface at over 400 mph. The lower part was expected to penetrate as far as 2 feet into the soil; the upper part of the probe was expected to stay on the surface in order to radio data to the Mars Global Surveyor spacecraft, which would then send the data to Earth. Unfortunately, there was no signal received from these probes. On December 7, 1999, project manager Sarah Gavit said she couldn't envision any failure scenario in which the batteries could still hold a charge after four days on Mars.

Mars Polar Lander and Mars Climate Explorer

Faster, Better, Cheaper Philosophy
Daniel S. Goldin became NASA administrator in 1992. He had a goal to reduce the cost of planetary missions all the way down to $150 million. He challenged JPL to adapt itself to his new "faster, better, cheaper" philosophy in a 1992 speech. Pathfinder had accomplished its mission for a total cost of $265 million, but this figure did not include the savings achieved from using spare Cassini parts, and other cost savings. JPL and its system contractor, Lockheed Martin, were then challenged to do the Mars Surveyor program, consisting of two missions, for $285 million.

The First Disaster
On December 11, 1998, the Mars Surveyor program launched the first of its two missions. Mars Climate Orbiter was designed to function as an interplanetary weather satellite and a communications relay for Mars Polar Lander. The orbiter carried two science instruments: a copy of an atmospheric sounder on the Mars Observer spacecraft lost in 1993, and a new, lightweight color imager combining wide- and medium-angle cameras. Mars Climate Orbiter was lost on arrival September 23, 1999. Engineers concluded that the spacecraft entered the planet's atmosphere too low and probably burned up.

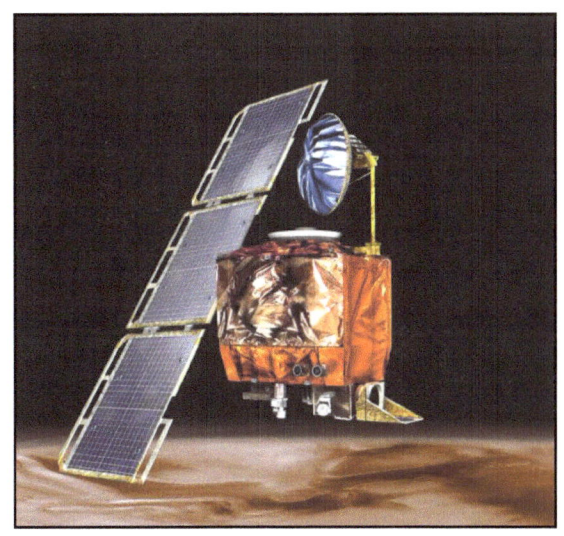

JPL knew within hours what had happened. A computer program in the ground control system had been written to use English Engineering Units (feet, pounds, etc.) instead of metric (meters, kilograms, etc.). This had caused the spacecraft to be just far enough off course to hit the atmosphere while going into orbit.

The Second Disaster

The second Mars Surveyor program mission, the Mars Polar Lander, was already on its way to Mars when the Orbiter was lost. The Mars Polar Lander launched on January 3, 1999 (see picture to the left).

Its target was the south polar region of Mars. It used a Viking-like lander equipped with a landing rocket and legs. In addition to the lander, it consisted of a cruise stage that would fly past Mars after releasing the lander, and two surface penetrator probes. An artist's concept of the lander is shown to the right.

The Polar Lander disappeared December 3, 1999, apparently crashing onto its landing site. The spacecraft had not been designed to provide telemetry during its descent so a definitive cause could not be found. But a test on duplicate hardware at Lockheed suggested that the most likely cause was a software fault that had shut off the descent rocket too early, causing the spacecraft to fall the last 65 yards to the surface.

Aftermath

After these losses, NASA overhauled the Mars program. The arbitrarily low cost of the Mars Surveyor Program led to cutting down on reviews, testing, and oversight to levels that were unsafe. The Mars Program responded by adding more money to future missions, and by reducing the objectives for the 2001 Mars mission to a single orbiter. The Faster/Better/Cheaper management philosophy had received a severe blow.

Collateral Impacts

Every project at JPL that launched after these failures was affected by a sudden increase in audits, reviews, and documentation requirements. This trend has continued from 2000 to this day. Cassini, Mars Global Surveyor, Pathfinder and Stardust all launched within their planned budgets. The ambitious Deep Space One mission launched within 6% of its planned budget. But very few JPL projects that started after 1999 have been able to stay within budget, partly due to escalating oversight efforts.

Stardust – A Comet's Tail

Stardust was dedicated to the exploration of a comet, and the first robotic mission designed to return extraterrestrial material from outside the orbit of the Moon.

The spacecraft was built for JPL by Lockheed Martin Aeronautics. It is shown in the image to the left.

Stardust was launched on February 7, 1999, from Cape Canaveral Air Station, Florida, aboard a Delta II rocket.

The primary goal of Stardust was to collect dust and carbon-based samples during its closest encounter with Comet Wild 2 (pronounced "Vilt 2" after the name of its Swiss discoverer), which took place in January 2004, after nearly four years of space travel.

The image to the right shows mission operations at the JPL Space Flight Operation Facility, at the time of the encounter.

A capsule containing aerogel was used to capture dust particles. This silica-based material was inserted within the Aerogel Collector Grid, which is similar to a large tennis racket. The image below shows an example of a piece of dust from the Comet, as it was captured within the aerogel.

Additionally, the Stardust spacecraft brought back samples of interstellar dust. These materials are believed to consist of ancient pre-solar interstellar grains that include remnants from the formation of the Solar System.

In order to meet up with comet Wild 2, the spacecraft made three loops around the Sun. On the second loop, its trajectory intersected the comet.

Stardust returned its aerogel capsule January 15, 2006, by parachuting a reentry capsule weighing approximately 125 pounds to a location in Utah. The picture below shows the capsule coming into Earth's atmosphere.

The picture to the left is the capsule on the ground. The picture to the right is a heart shaped piece of comet dust.

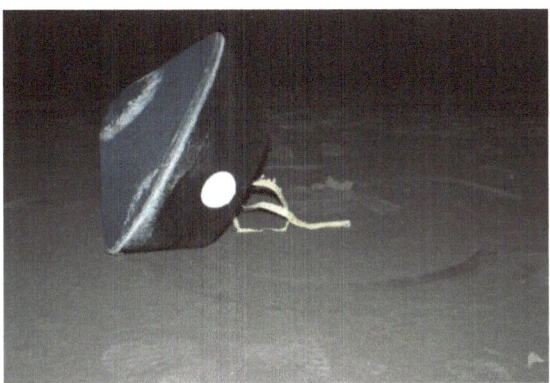
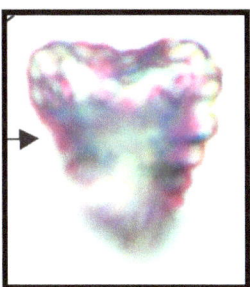

Comet Wild 2 is a collection of materials that probably came from all regions of the young solar system and thus it has turned out to be wonderful "time capsule". Hundreds of scientists around the world have worked on these samples and the first results from these studies were presented in the December 15, 2006 issue of Science magazine. Having samples from the edge of the solar system has provided a fabulous way to explore the early solar system and test ideas for its origin.

Ken Atkins managed Stardust during operations. Tom Duxbury was the project manager during operations. Lockheed Martin built the spacecraft. Instruments came from JPL (dust collector); Germany (Cometary and Interstellar Dust Analyzer); and the University of Chicago (Dust Flux Monitor).

The spacecraft continues to fly in deep space, and is now being used in a second mission, called EPOXI, which will be described later.

Wide-field Infrared Explorer – The High Cost of Cheaper Spaceflight

The Wide-field Infrared Explorer (WIRE) was a small satellite carrying a cryogenically cooled infrared telescope designed to study starburst galaxies -- vast clouds of molecular gas cradling the sites of newborn stars. It was developed under NASA's Small Explorer Program, at the height of the "Faster, Better, Cheaper" philosophy of space missions.

WIRE was intended to have a four-month primary mission, at which point the coolant would be fully consumed.

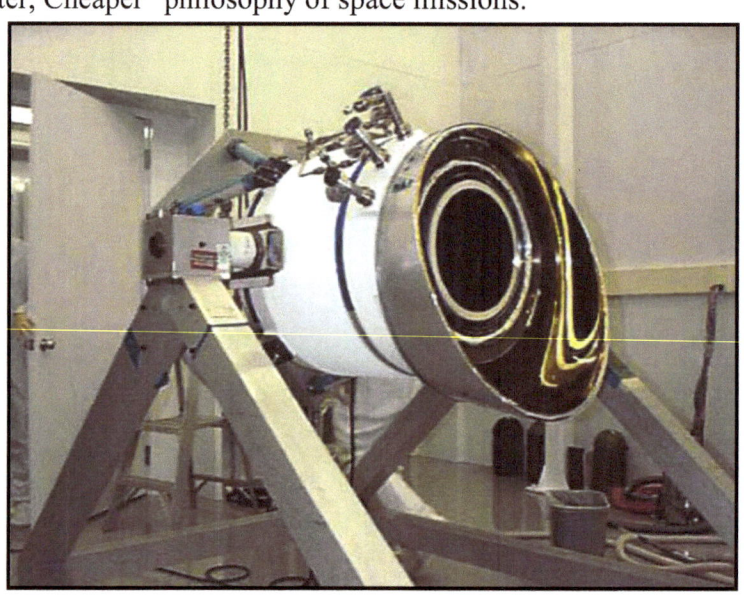

The infrared telescope (shown to the right) was built for JPL by the Space Dynamics Laboratory of Utah State University in Logan, Utah. The WIRE spacecraft was designed and built by the Small Explorer project team at NASA's Goddard Space Flight Center.

WIRE was launched on May 4, 1999. The satellite was placed in orbit around Earth at an altitude of 335 miles. Due to a spacecraft equipment malfunction, the telescope's coolant was rapidly depleted shortly after launch, leaving the science instrument unusable for its original science objective.

This meant that over the course of about a year, four JPL flight projects (Deep Space 2; Mars Climate Orbiter; Mars Polar Lander; and WIRE) failed. From this point forward, the faster, better, cheaper approach to missions would be replaced by progressive increases in regulation and oversight – a trend that has continued since 2000 to the date this book was written in 2009.

Quikscat – Measuring Ocean Winds

The ocean-observing Quick Scatterometer (Quikscat) satellite launches with the Seawinds instrument onboard on June 19, 1999. The mission was a quick turn-around replacement to Japan's Midori satellite, which carried the NASA Scatterometer (NSCAT), which lost power in June 1997.

Quikscat circles the Earth every 101 minutes. Its Seawinds instrument senses ripples in the ocean's surface caused by winds. These ripples are analyzed by scientists who compute the winds' speed and direction. The artist concept below shows the satellite in orbit.

The Seawinds instrument acquires vastly more observations of surface wind velocity each day than can ships and buoys. It can get data from storms that no ship would survive. The image to the right is from Hurricane Dora.

Jim Graf managed this project during implementation. Robert Gaston manages this project during operations. Ball Aerospace built the spacecraft. JPL built the instrument.

A Back Story

Quikscat is the most rapid deployment of a satellite in recent JPL history. It was originally designed to last for two years, as an emergency replacement for the lost Midori satellite. Quikscat is still operating today, ten years later. At least two replacement missions have been planned at JPL, but have not been built because this satellite just keeps on going.

97

MISR and ASTER on NASA's Terra Satellite

NASA's Terra satellite put two JPL-managed imaging instruments in orbit around Earth.

ASTER

The Advanced Spaceborne Thermal Emission and Reflection Radiometer (ASTER) obtains high-resolution global, regional and local images of Earth in 14 color bands. ASTER is a cooperative effort between NASA, Japan's Ministry of Economy, Trade and Industry (METI) and Japan's Earth Remote Sensing Data Analysis Center. ASTER is being used to obtain detailed maps of land surface temperature, reflectance and elevation.

Ice caps can be precisely monitored by this instrument.

The image to the right was acquired on May 2, 2000 over the North Patagonia Ice Sheet, Chile. The image covers 36 x 30 km.

The false color composite displays vegetation in red. The image dramatically shows a single large glacier, covered with crevasses. A semi-circular terminal moraine indicates that the glacier was once more extensive than at present.

ASTER monitors fires. The image to the left shows areas devastated by bushfires in Victoria, on Australia's East Coast. The image combines visible light with near-infrared light, and although the resulting false-color image doesn't look like a natural photo, it makes the burned areas (charcoal-brown) stand out better from unburned vegetation (red) and areas where vegetation is naturally sparse or dormant (beige).

Image credit: NASA/Jesse Allen

ASTER also monitors volcanoes around the world, with images stored in the ASTER volcano archive.

The image to the right was an "urgent acquisition". Asama Volcano on the Japanese island of Honshu showed signs of unrest starting in late January 2009. ASTER acquired this image of the volcano on February 7, 2009. In this false-color image, red indicates vegetation, gray-beige indicates bare rock, and white indicates snow or vapor. A generous cap of snow surrounds the volcano's summit in this wintertime shot, but a dark gray-brown stain appears south of the summit, thanks to a recent ash eruption. A white plume of vapor also appears near the summit

Dave Nichols was Project Manager during implementation. Bjorn Eng is now the Project Manager during operations.

MISR

The Multi-angle Imaging SpectroRadiometer, unlike any other instrument ever flown in space, views Earth from nine widely-spaced angles as the Terra satellite glides above Earth.

Since other instruments only see Earth from one perspective, this instrument greatly enhances imaging capabilities.

This is the "science part" of the MISR instrument, which includes the cameras and calibration equipment. The photograph was taken in October 1996, as MISR was being assembled. Subsequently, the parts that supply power, communications, and temperature control were added. The entire package was then encased in a protective housing, which was covered with highly reflecting thermal blankets.

MISR provides new types of information for scientists studying Earth's climate.

MISR monitors the monthly, seasonal, and long-term trends in:

- The amount and type of atmospheric particles (aerosols), including those formed by natural sources and by human activities
- The amounts, types, and heights of clouds
- The distribution of land surface cover, including vegetation canopy structure

Aerosols tend to cool the surface below them. This can counter-act global warming to some degree, but careful measurements over an extended period of time are needed to accurately predict how much cooling can take place.

The image to the right shows large fires that were active in Alaska and the Yukon Territory from mid-June to mid-July, 2004. Thick smoke particles filled the air during these fires, prompting Alaskan officials to issue air quality warnings. Some of the smoke from these fires was detected as far away as New Hampshire. These images were captured on June 30th by MISR. Here, MISR distinguishes clouds 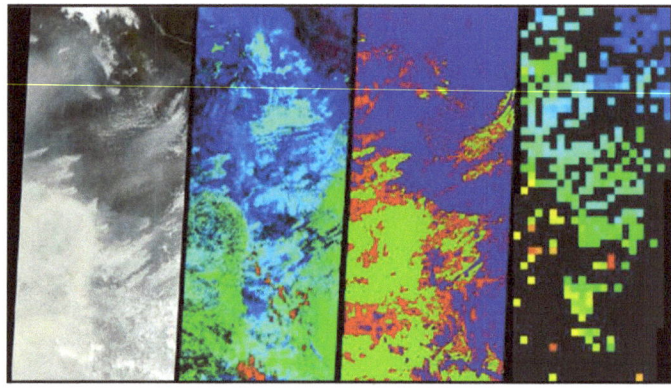 from smoke and retrieves heights and optical depths for the smoke -- information which will help to improve models of how smoke aerosols are transported.

Terry Reilly and Graham Bothwell are former managers of MISR. Earl Hansen is currently the Project Manager during operations.

The ACRIM Satellite – Measuring Sunlight

The Active Cavity Radiometer Irradiance Monitor Satellite (AcrimSat) is designed to monitor the total amount of the Sun's energy reaching Earth.

It is this energy, called total solar irradiance, that creates the winds, heats the land and drives ocean currents.

Some scientists theorize that a significant fraction of Earth's global warming may be solar in origin due to small increases in the Sun's total energy output since the last century. By measuring incoming solar radiation, climatologists will be able to improve their predictions of climate change and global warming.

AcrimSat was launched December 20, 1999, as a secondary payload on a Delta II rocket from California's Vandenberg Air Force Base. AcrimSat circles Earth from a polar orbit at an altitude of about 425 miles. It is operating to this day.

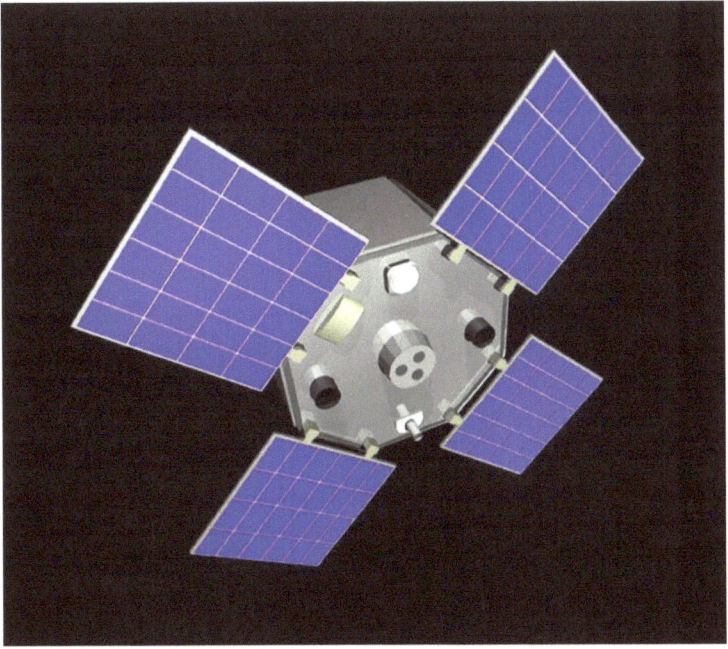

Ronald J. Zenone managed this project during implementation. Roger Helezon is now managing this project during operations. Orbital Sciences Corporation built the spacecraft.

Mars Odyssey – Exploring Mars

Mars Odyssey is an orbiter that carries a spectrometer designed to identify and quantify substances as deep as 1 meter below the Martian surface. This data would be used by scientists to improve our understanding of the planet's climate and geologic history.

Mars Odyssey launched on April 7, 2001, and arrived at Mars on October 24, 2001. An artist's conception of Mars Odyssey at Mars is shown below.

Odyssey used aerobraking to bring it closer to Mars with each orbit. By using the atmosphere of Mars to slow down the spacecraft in its orbit rather than firing its engine or thrusters, Odyssey saved more than 440 pounds of propellant. Odyssey is still operating to this day.

An important finding was announced on May 28, 2002:

> "surprised scientists have found enormous quantities of buried treasure lying just under the surface of Mars-enough water ice to fill Lake Michigan twice over. And that may just be the tip of the iceberg."

The Odyssey orbiter has also provided a communications relay for the Mars Exploration Rovers, Spirit and Opportunity, transmitting 85% of the data from the rovers to Earth.

George Pace was the original Project Manager. Roger Gibbs and Phil Varghese have also managed this project. The spacecraft was built by Lockheed Martin.

More on JPL Management

Charles Elachi was appointed Director of the Jet Propulsion Laboratory in May 2001, and he continues to lead JPL to this day.

In October, 2006 Elachi was honored as one of "America's Best Leaders" by the U.S. News & World Report, in collaboration with the Center for Public Leadership at Harvard University's John F. Kennedy School of Government.

JPL had experienced a flood of change initiatives in the 1990s. Changes and reorganizations have continued in this decade, but at a more measured pace.

Elachi brings a spirit of great optimism and energy to the Laboratory. The number of active missions at JPL has continued to increase under his leadership, as can be seen in the figure below. The green bars show approximately how many missions are starting, the red bars show how many missions are being prepared for launch, and the blue bars show the number of missions that are operating in outer space. This figure does not include instrument projects, or proposal activities. Numbers were counted at the beginning of each year.

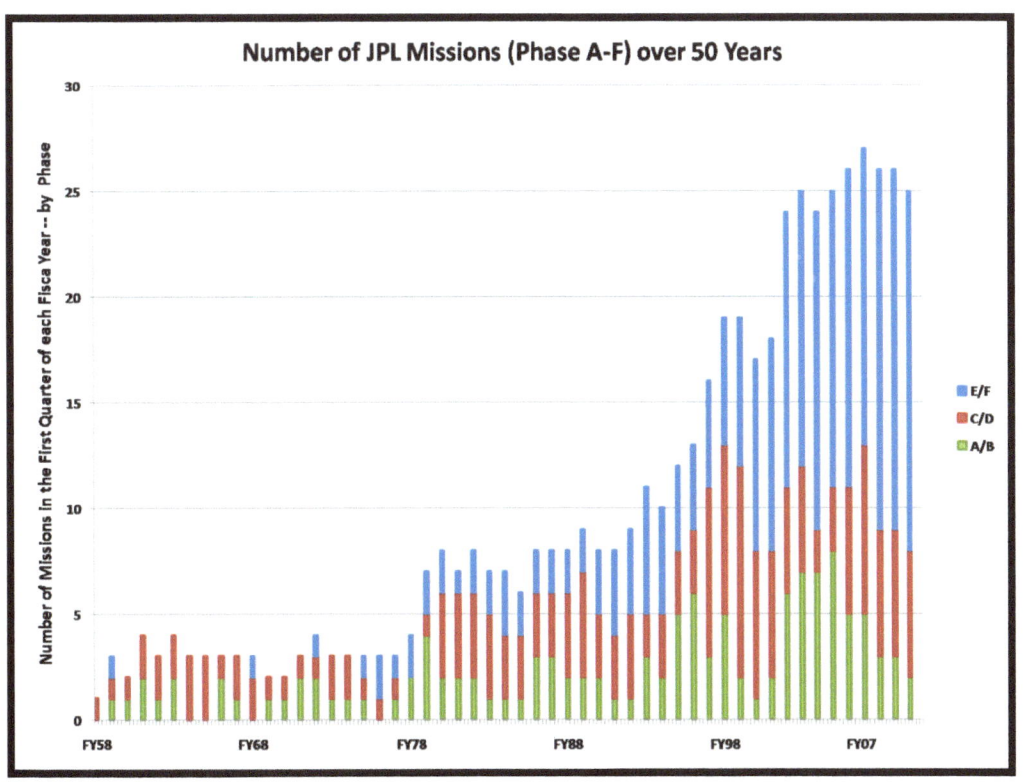

Genesis – Catching Pieces of the Sun

Most scientists believe that the solar system was formed when a cloud of gas and dust nearly 4-1/2 billion years ago, forming the Sun, planets, comets and asteroids. Exactly how that transformation took place both intrigues and mystifies scientists.

The Genesis mission collected samples of the solar wind, material flowing outward from the Sun. Comparing them with known compositions of the planets will help in the effort to understand our cosmic origins.

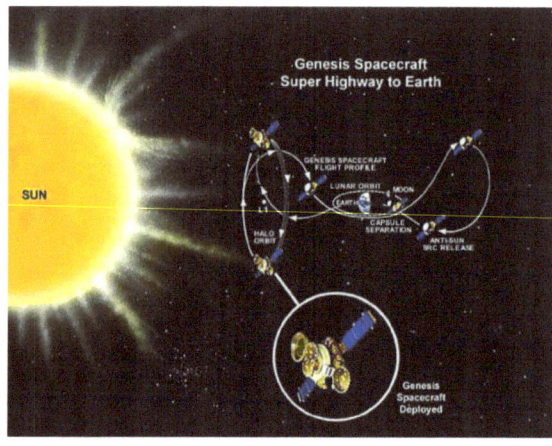

Following launch on August 8, 2001, the Genesis spacecraft headed toward an orbit around L1, a point between Earth and the Sun where the gravity of both bodies is balanced. This is about 1 million miles from Earth. See picture at right.

Genesis unfurled its collector arrays (see below) and collected particles of the solar wind that embedded themselves in specially designed high purity wafers.

In 2004, Genesis returned the samples to Earth. This was the first sample return since the Apollo program, in 1972.

Sample return was designed to be accomplished by deploying a return capsule which would enter Earth's atmosphere at high speed which was then supposed to release a series of parachutes over the Utah Test & Training Range, and then was supposed to be caught by helicopters before it hit the ground (shown at right).

Crash!

The sample return capsule's parachutes did not deploy, and the capsule struck the Utah Test & Training Range at a speed of 193 miles per hour. The capsule was cracked.

The capsule was recovered, and it was found that most collection cells were undamaged, and it would be possible to clean them and still recover atoms that came from the sun.

The sample analysis process has continued to this day. It had been assumed that Earth and the sun would have similar ratios of isotopes of oxygen and other elements. Genesis shows that this is not the case, and that there are significant differences between the composition of the Sun and the planets that have been carefully analyzed at the atomic level (Earth, Mars, and Earth's moon).

Chet Sasaki managed Genesis up through launch. Don Sweetnam managed Genesis during flight operations. Lockheed Martin built the spacecraft. The Johnson Space Center manages the returned material, and has distributed samples to twenty eight laboratories to date.

Jason – Continuing the TOPEX Legacy

The objective of the Jason mission is to continue the TOPEX/Poseidon mission. Jason provides high accuracy measurements of sea surface topography and continues the program of long-term observations of ocean circulation.

Scientifically, this means obtaining sea surface height measurements with a global accuracy of better than 2 inches.

Jason launched in December, 2001. It completed its primary mission in December, 2004. However, it is in excellent working order and continues to operate to this day.

The latest images from Jason indicate that California is currently undergoing a La Niña weather condition (see image to the right).

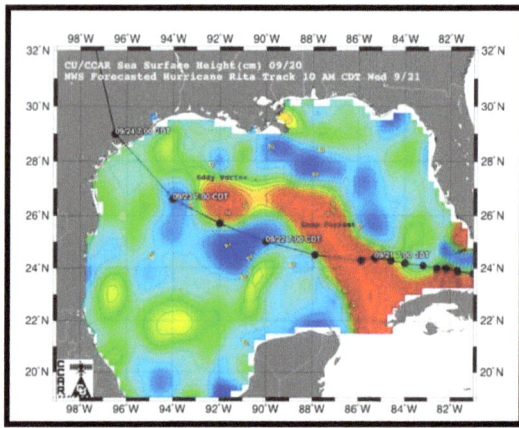

The image to the left shows Hurricane Rita. It combines several sources of data, including Jason. Jason contributes the sea surface height map of the Gulf of Mexico (the Florida peninsula is on the right and the Texas-Mexico Gulf Coast on the left). Red indicates a strong circulation of much warmer waters, which can feed energy to a hurricane. This area stands 13 to 23 inches higher than the surrounding waters of the Gulf.

The Project Manager during implementation was Gary Kunstmann, and during operations it is Mark Fujishin. JPL provided instruments, the French Space Agency provided the rest of the flight system.

GRACE – Precision Gravity Measurement

The primary goal of the GRACE mission is to obtain accurate global and high-resolution models for both the static and the time variable components of the Earth's gravity field approximately every 30 days. These gravity maps are up to 1,000 times more accurate than those previously available.

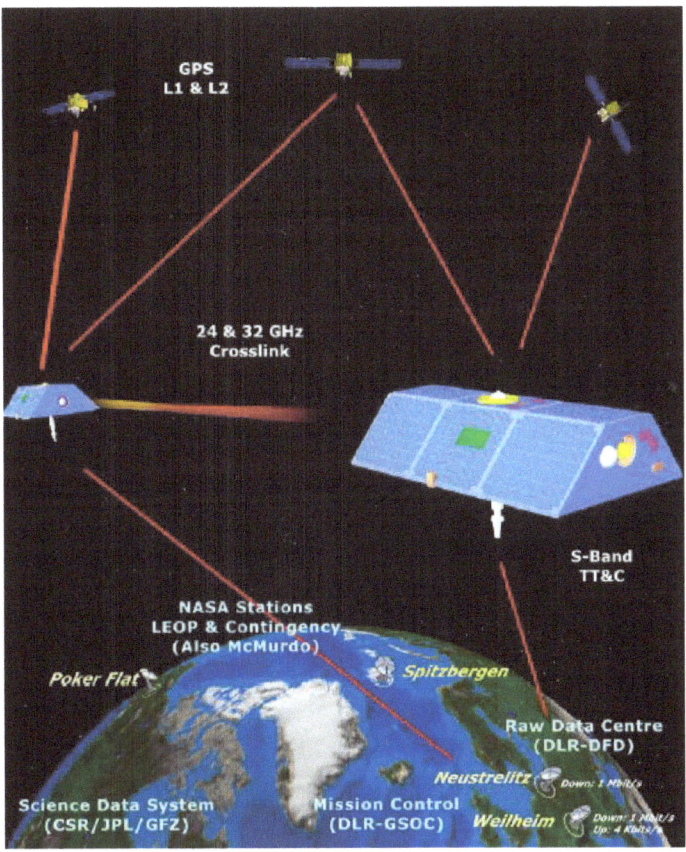

This is accomplished by obtaining very accurate measurement of the altitudes of two spacecraft using the Global Positioning System and ultra-accurate (within 10 microns) measurement of the distance between the two spacecraft. The mission is able to measure changes in sea-floor pressure and show how the mass of the oceans change. It also measures and monitors ice sheets and changes in the storage of water and snow on the continents.

GRACE launched from Plesetsk, Russia on March 17, 2002, and has continued its mission to this day.

High accuracy gravity maps substantially improve the accuracy of many techniques used by oceanographers, hydrologists, glaciologists, geologists and other scientists to study phenomena that influence climate. These phenomena range from shallow and deep ocean currents, water movement on and beneath Earth's surface, and the movement and changing mass of ice sheets, to sea-level heights, sea-level rise and changes in the structure of the solid Earth

For example, changes in the mass of water in an area can produce detectable changes in Earth's average gravity field at that location. GRACE measurements allow scientists to derive changes in water storage over large regions (greater than 200,000 square kilometers). The panel of images below shows how water storage in the four major sub-basins of the United States' Mississippi River basin differed from average in January (top row) and July (bottom row) 2005. The "GRACE data plus models" includes a limited amount of on-site measurement, which increases the accuracy of the overall map.

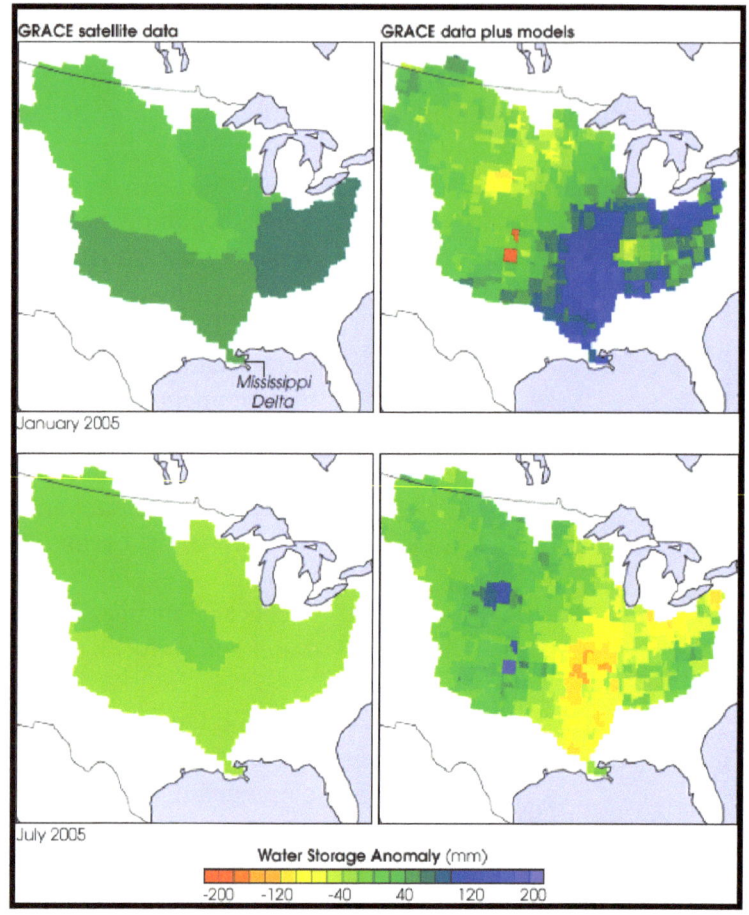

GRACE was managed by Ab Davis during implementation, and by Joe Beerer during operations. Germany provided the spacecraft, Russia provided the launch vehicle, and JPL produced the instrument.

A Back Story
The "Space Race" was very important in the 1960s and 1970s. But GRACE was a highly cooperative effort between the US, German, and Russian space agencies.

Atmospheric Infrared Sounder (AIRS)

The Atmospheric Infrared Sounder is an instrument onboard NASA's Aqua satellite. The term "sounder" in the instrument's name means measurements are taken at various elevations in Earth's atmosphere. AIRS was designed and built by Lockheed Infrared Imaging Systems under contract with JPL. The Aqua satellite mission is managed by the Goddard Space Flight Center.

The AIRS image below shows Katrina when it was on the Mississippi-Tennessee border. It shows the area of most intense precipitation, by measuring the temperature of the cloud tops

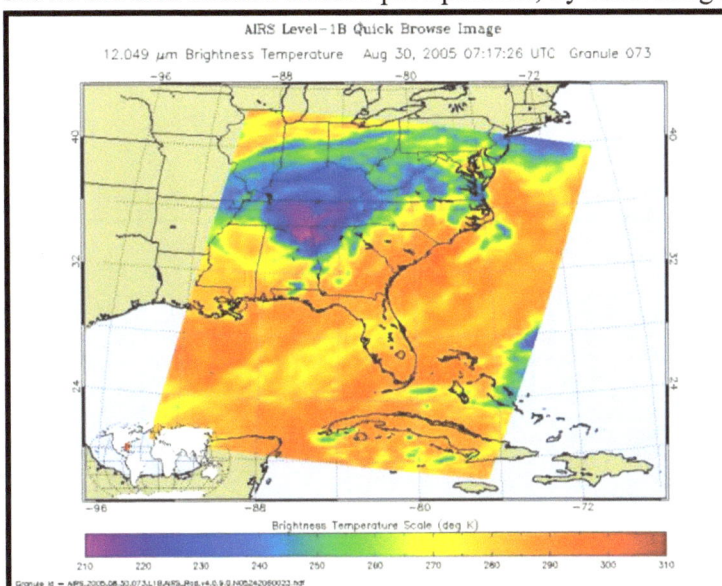

(or the surface of the Earth in cloud-free regions). The lowest temperatures are associated with high, cold cloud tops that make up the top of the hurricane. The infrared signal does not penetrate through clouds, so the purple color indicates the cool cloud tops of the storm. In cloud-free areas, the infrared signal is retrieved at the Earth's surface, revealing warmer temperatures. Cooler areas are pushing to purple and warmer areas are pushing to red.

AIRS provides maps of carbon monoxide in the atmosphere, as shown in the image to the right.

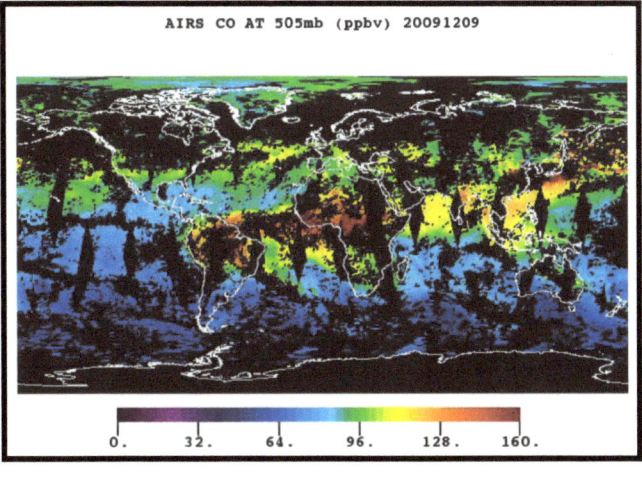

AIRS also provides maps of atmospheric temperature during the day and night. The image to the right shows the day temperature.

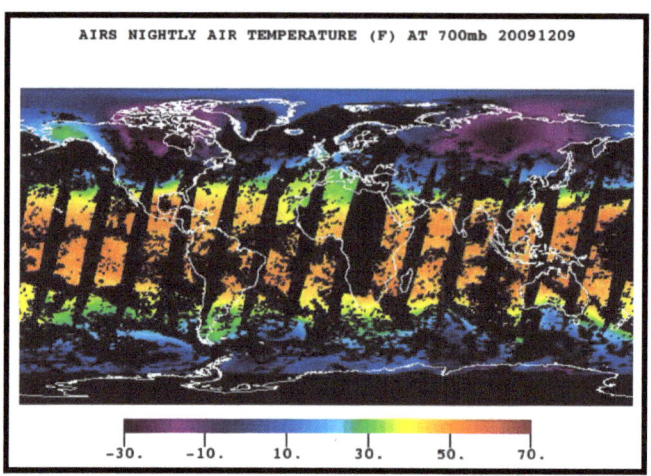

The map to the right shows ozone level totals, from the Earth's surface to the top of Earth's atmosphere.

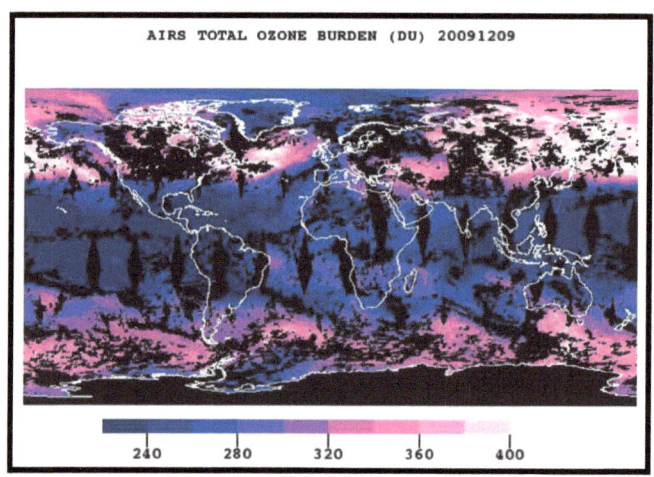

The project manager during implementation was Dr. Avi Karnik. The manager during operations is Tom Pagano. AIRS was built for JPL by Lockheed Infrared Imaging Systems.

Results
Data is used to better understand weather and climate. There have been hundreds of scientific investigations that led to publications, using AIRS data.

AIRS data is also used by the National Weather Service and the National Oceanic and Atmospheric Administration to improve the accuracy of their weather and climate models.

GALEX – the Ultraviolet Explorer of the Universe

The Galaxy Evolution Explorer (GALEX) is an orbiting space telescope that makes observations at ultraviolet wavelengths to measure the history of star formation in the universe across about 10 billion years of cosmic history. It can be seen below, as it is being readied for launch.

The spacecraft's mission is to observe hundreds of thousands of galaxies, with the goal of determining how far away each galaxy is from Earth and how fast stars are forming in each galaxy.

A galaxy's ultraviolet brightness tells us how fast its stars are forming. Mission scientists look for stars that have recently formed. These are the most massive stars, so hot that they shine in ultraviolet wavelengths.

Galex was managed by Jim Fanson during implementation and by Kerry Erickson during operations. Participation included the California Institute of Technology (Principal investigator); Orbital Sciences Corporation (Spacecraft); University of California at Berkeley; Yonsei University, in Seoul, Korea; Johns Hopkins University; and the Laboratoire d'Astrophysique de Marseille, in France. JPL built the instrument.

GALEX was selected in 1997 and went into development with a faster, better, cheaper implementation approach, consistent with similar missions at that time. A string of NASA mission failures (WIRE, Lewis, MPL and MCO) caused a reassessment of acceptable risk, leading to changes to the GALEX implementation approach, schedule delays, and cost increases.

The mission is a technical triumph. GALEX is producing the first complete map of galaxies under construction. The imaging survey will identify 1,000,000 galaxies. The spectroscopic surveys will examine 100,000 galaxies. One recent example is an April 16, 2008 image from NASA's Galaxy Evolution Explorer which shows baby stars sprouting in a relatively desolate region of space more than 100,000 light-years from the galaxy's bustling center.

The image below is a composite of ultraviolet data from the Galaxy Evolution Explorer and radio data from the National Science Foundation's Very Large Array in New Mexico, shows the Southern Pinwheel galaxy, also known simply as M83. M83 is located 15 million light-years away in the southern constellation Hydra. In this image, far-ultraviolet light is blue, near-ultraviolet light is green and radio emission at a wavelength of 21 centimeters is red.

In the new view, the main spiral, or stellar, disk of M83 looks like a pink and blue pinwheel, while its outer arms appear to flap away from the galaxy like giant red streamers. It is within these so-called extended galaxy arms that, to the surprise of astronomers, new stars are forming. This came as a surprise to astronomers because the outlying regions of a galaxy are assumed to be relatively barren and lack high concentrations of the ingredients needed for stars to form.

The astronomers speculate that the young stars seen far out in M83 could have formed under conditions resembling those of the early universe, a time when space was not yet enriched with dust and heavier elements.

Image Credit: NASA/JPL-Caltech/VLA/MPIA.

"Even with today's most powerful telescopes, it is extremely difficult to study the first generation of star formation. These new observations provide a unique opportunity to study how early generation stars might have formed," said co-investigator Mark Seibert of the Observatories of the Carnegie Institution of Washington in Pasadena.

Mars Exploration Rover – The Mars Rovers that Keep On Going

Mars Exploration Rover was a high risk mission due to its complexity and the relatively short amount of time available to bring it off. Two thirds of all previous missions to Mars (including Soviet missions) had previously failed.

It takes three and a half years to proceed from a mission concept to an actual spacecraft on the launch pad, if it is a relatively easy mission. And it normally takes five or more years to get to the launch pad with a complex mission that uses new technology.

The Mars Program was in disarray after the MCO and MPL failures in 1999. Odyssey had already started, and would launch in April, 2001, but there was uncertainty over what to do as the next step in Mars exploration. Launch windows are established by the relative position of Earth and Mars. Odyssey launched in 2001. The next window would be in June, 2003. A decision on how to proceed with the 2003 launch opportunity would normally have occurred in 1999 or earlier. But the MCO and MPL failures and the failure review process captured so much attention that a decision on how to proceed for the 2003 window was not made until mid-2000, leaving just three years to do the job.

The rovers used for this mission were a new technology, and were much heavier (400 pounds) and more capable than the Pathfinder rover. This made it necessary to redesign the entry, descent and landing approach that had worked previously.

The image to the left shows parachute testing.

The combination of new technology and an abbreviated schedule posed huge technical and management challenges to JPL.

The Mars Exploration Rover project was a top priority at JPL from its beginning through to its launch. Cost management was sacrificed to meet extremely tight schedules and to mitigate as much of the technical risk as possible. People working on this project routinely spent evenings and weekends on the job to make up schedule time. The integration and test phase started with two shifts a day, and for much of this phase the work continued around the clock. MER was managed by Pete Theisinger until February 2004, followed by Richard Cook, Jim Erickson, and (currently) John Callas during operations. The spacecraft were built at JPL.

Each rover has five scientific instruments and a rock abrading device. The Panoramic Camera and the Miniature Thermal Emission Spectrometer are located on the large mast shown on the front of the rover. The camera was supplied by JPL and the spectrometer was supplied by Arizona State University. The payload also includes magnetic targets, provided by the Niels Bohr Institute in Copenhagen, Denmark, that will collect magnetic dust for further study by the science instruments.

The Rock Abrasion Tool is located on a robotic arm that can be deployed to study rocks and soil.(In this view, the robotic arm is tucked under the front of the rover.) The tool, provided by Honeybee Robotics Ltd., New York, N.Y., can grind away the outer surfaces of rocks, which may be dusty and weathered, allowing the science instruments to determine the nature of rock interiors. The three instruments that study abraded rocks are a Mossbauer Spectrometer, provided by the Johannes Gutenberg- University at Mainz, Germany; an Alpha-Proton X-ray Spectrometer provided by Max Planck Institute for Chemistry, also in Mainz, Germany; and a Microscopic Imager, supplied by JPL.

The Mars Exploration Rover mission had two launches. Spirit launched on June 10, 2003; Opportunity launched on July 7, 2003.

Spirit and Opportunity were landed at sites on opposite sides of Mars. The landing sites are Gusev Crater, a possible former lake in a giant impact crater, and Meridiani Planum, where mineral deposits (hematite) suggest Mars had a wet past. The image below is the Bonneville crater, taken by Spirit.

Primary among the mission's scientific goals was to search for and characterize a wide range of rocks and soils that hold clues to past water activity on Mars. The image below is Cape Saint Mary, taken by Opportunity.

The image to the left shows Opportunity at work, analyzing a sample inside Victoria Crater, on Mars.

The primary mission of each Rover was planned for 3 months in duration. JPL celebrated their first five years of operation in January, 2009. Both rovers still operate.

One reason the rovers continue to operate is the test facility that is used at JPL to this day. The picture to the right above shows an engineering model that is being used to diagnose problems that arise with the rovers on Mars.

MARSIS on the Mars Express

Not only did JPL launch MER, it also contributed the MARSIS radar to a European mission to Mars, the Mars Express.

The Mars Advanced Radar for Subsurface and Ionospheric Sounding (MARSIS) is a subsurface radar with a 130-foot antenna on the European Space Agency's Mars Express orbiter.

MARSIS can look for water from the Martian surface down to about 3 miles below the surface. It also characterizes surface elevation, roughness, and radar reflectivity of the planet and studies the interaction of the atmosphere and solar wind in the red planet's ionosphere.

The best ground-penetrating studies are made during the night when the Martian ionosphere is least active and when the spacecraft is less than 800 kilometers from the Martian surface, a condition that occurs for 26 minutes during each orbit.

During the day, sunlight ionizes the upper atmosphere (charges it up electrically) and long wavelength radio waves bounce off it. Those that are reflected from the ionosphere can reveal its structure.

Mars Express launched on June 2, 2003. Mars orbit insertion occurred on December 24, 2003. Antenna deployment occurred after Mars orbit insertion.

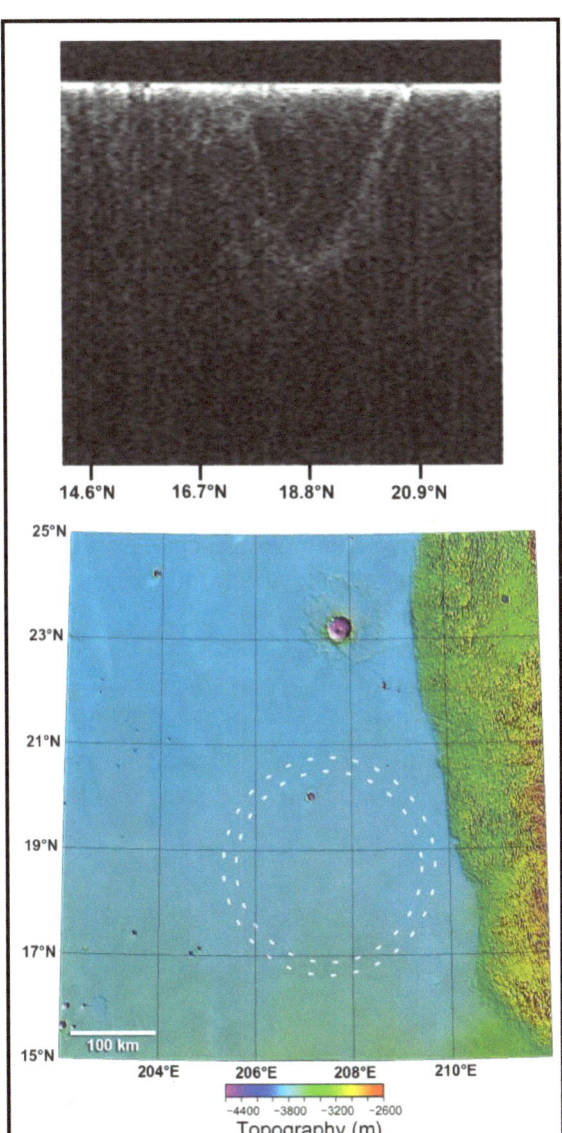

The top image to the left is a radargram presenting data collected by the MARSIS during the 1,886th orbit of the European Space Agency's Mars Express orbiter.

It shows parabolic-shaped echoes from the rim walls of a buried impact basin.

In the lower image, parabolic echoes project to circular arcs on the surface and indicate the location of a 130-mile-diameter impact basin buried by young lava flows.

Image Credit: ASI/NASA/ESA/JPL-Caltech/Univ. of Rome

The MARSIS project was managed by Richard Horttor of JPL during implementation, and by Thomas Thompson during operations.

A Sad Note

The European Space Agency attempted to put the Beagle lander onto the surface of Mars on Dec 24, 2003, but the attempt was unsuccessful.

Historically, Mars has been a tough target. A majority of Mars missions from the Soviet Union, the USA, and Europe have failed. But our collective success rate has improved over time.

Spitzer Space Telescope – The Great Infrared Observatory

The Spitzer Space Telescope was launched into space on 25 August 2003. Spitzer obtains images and spectra by detecting the infrared energy, or heat, radiated by objects in space. Most of this infrared radiation is blocked by the Earth's atmosphere and cannot be observed from the ground.

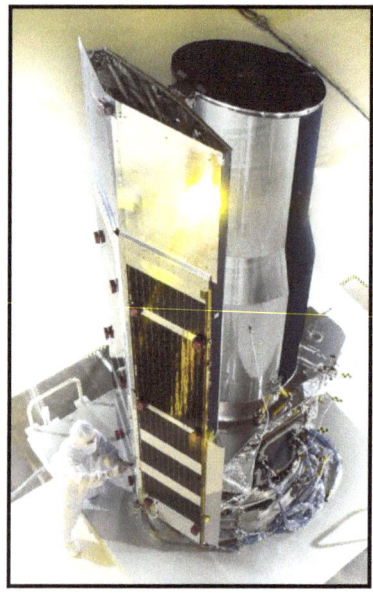

Spitzer has a 0.85-meter telescope and three cryogenically-cooled science instruments. It is the largest infrared telescope ever launched into space.

Because infrared is primarily heat radiation, the telescope must be cooled to near absolute zero (-459 degrees Fahrenheit or -273 degrees Celsius) so that it can observe infrared signals from space without interference from the telescope's own heat. Also, the telescope must be protected from the heat of the Sun and the infrared radiation put out by the Earth.

To do this, Spitzer carries a solar shield and was launched into an Earth-trailing solar orbit. This unique orbit places Spitzer far enough away from the Earth to allow the telescope to cool rapidly without having to carry large amounts of coolant. This innovative approach has significantly reduced the cost of the mission.

Many areas of space are filled with vast, dense clouds of gas and dust which block our view. Infrared light can penetrate these clouds however, allowing us to peer into regions of star formation, the centers of galaxies, and into newly forming planetary systems.

This false-color composite image of the Stephan's Quintet galaxy cluster clearly shows one of the largest shock waves ever seen (the green arc). The wave was produced by one galaxy falling toward another at speeds of more than one million miles per hour.

Infrared also brings us information about the cooler objects in space, such as smaller stars which are too dim to be detected by their visible light, extrasolar planets, and giant molecular clouds. Also, many molecules in space, including organic molecules, have their unique signatures in the infrared.

A star's spectacular death in the constellation Taurus was observed on Earth as the supernova of 1054 A.D.

Now, almost a thousand years later, a super-dense neutron star left behind by the stellar death is spewing out a blizzard of extremely high-energy particles into the expanding debris field known as the Crab Nebula.

Image Credit: NASA/JPL-Caltech/ESA/CXC/Univ. of Ariz./Univ. of Szeged

The galaxies below have taken on the form of a giant mask. The icy blue eyes are actually the cores of two merging galaxies, called NGC 2207 and IC 2163, and the mask is their spiral arms. The false-colored image consists of infrared data from NASA's Spitzer Space Telescope (red) and visible data from NASA's Hubble Space Telescope (blue/green).

Image Credit: NASA/JPL-Caltech/STScI/Vassar

This false-color composite image shown below is the Cartwheel galaxy as seen by the Galaxy Evolution Explorer (blue); the Hubble Space Telescope (green); the Spitzer Space Telescope (red); and the Chandra X-ray Observatory (purple).

Approximately 100 million years ago, a smaller galaxy plunged through the heart of Cartwheel galaxy, creating ripples of brief star formation. In this image, the first ripple appears as an ultraviolet-bright blue outer ring. The blue outer ring is so powerful in the Galaxy Evolution Explorer observations that it indicates the Cartwheel is one of the most powerful UV-emitting galaxies in the nearby universe. The blue color reveals to astronomers that associations of stars 5 to 20 times as massive 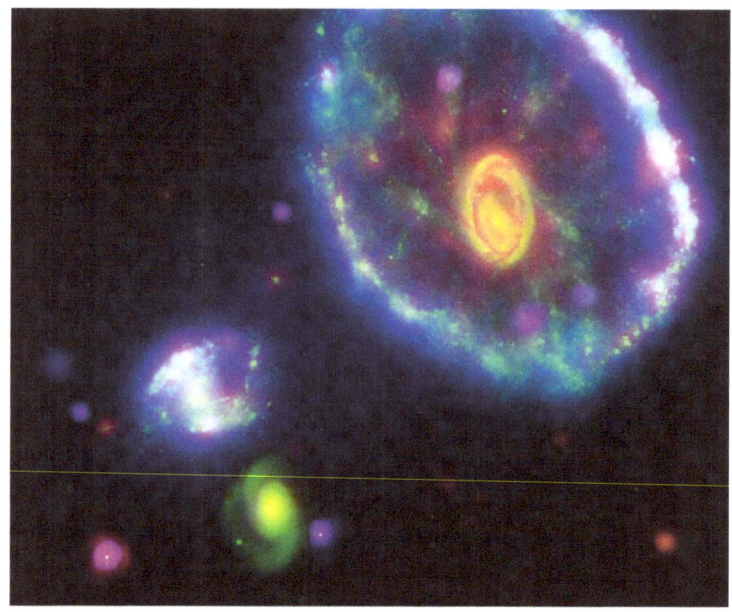 as our sun are forming in this region. The clumps of pink along the outer blue ring are regions where both X-rays and ultraviolet radiation are superimposed in the image. These X-ray point sources are very likely collections of binary star systems containing a black hole.

The yellow-orange inner ring and nucleus at the center of the galaxy result from the combination of visible and infrared light, which is stronger towards the center. This region of the galaxy represents the second ripple, or ring wave, created in the collision, but has much less star formation activity than the first (outer) ring wave. The wisps of red spread throughout the interior of the galaxy are organic molecules that have been illuminated by nearby low-level star formation. Meanwhile, the tints of green are less massive, older visible-light stars. Although astronomers have not identified exactly which galaxy collided with the Cartwheel, two of three candidate galaxies can be seen in this image to the bottom left of the ring, one as a neon blob and the other as a green spiral.

Larry Simmons and David Gallagher managed Spitzer to launch. Robert Wilson manages it now, during operations. Lockheed Martin built the spacecraft. Payload elements were built by the Smithsonian Astrophysical Observatory; Cornell University; the University of Arizona; and Ball Aerospace and Technology Corporation.

Microwave Limb Sounder on AURA – Studying Earth

There was an MLS instrument on the UARS mission several years earlier. Measurements of chlorofluorocarbons and other gases in Earth's atmosphere are important for studying changes in the Earth's ozone layer. A continuous record of data over many years is highly desirable for modeling long term trends.

The overall objective of MLS research is to improve our understanding of how Earth's atmosphere works. The unique MLS measurements of many stratospheric chemicals are essential for understanding ozone layer stability and possible threats to it.

Its measurements of upper tropospheric water vapor and cloud ice, which have historically been difficult, provide key data on feedback processes affecting climate change.

Its measurements of pollution in the upper troposphere, including in the presence of ice clouds that prevent measurement by other techniques, provide unique data on the long-range transport of pollution and its possible effects on climate.

The MLS instrument on AURA is shown to the right.

The Aura mission launched 15 July 2004.

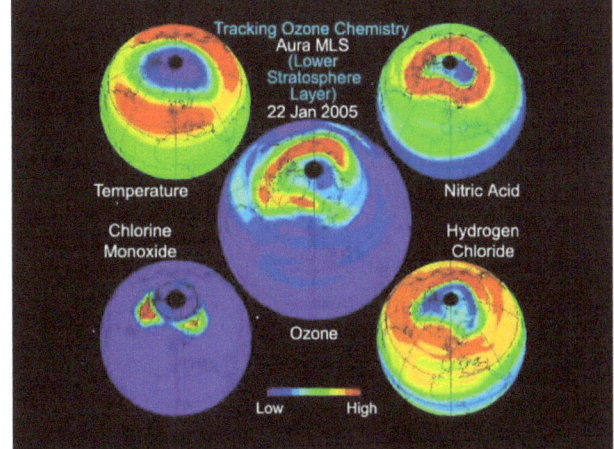

EOS MLS began full-up atmospheric science observations on 13 August 2004, with excellent performance to date in all portions of the instrument. MLS can map several gases in Earth's atmosphere, including ozone, as shown in the image above. The ozone hole is clearly identified.

MLS also measures atmospheric water, nitric acid, hydrochloric acid, Chlorine monoxide (same as MLS on UARS), Bromine monoxide, Nitrous oxide, volcanic sulfur dioxide, and other chemicals that can affect ozone in the stratosphere.

The image to the right shows Nitrous oxide, and shows its relationship to ozone formation.

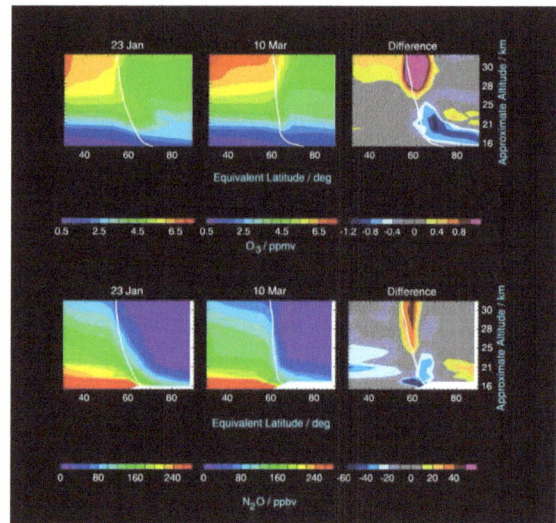

The image below shows Carbon Monoxide as a function of altitude, over time. This is the kind of data that enables scientists to conduct basic research on Earth's evolving atmosphere.

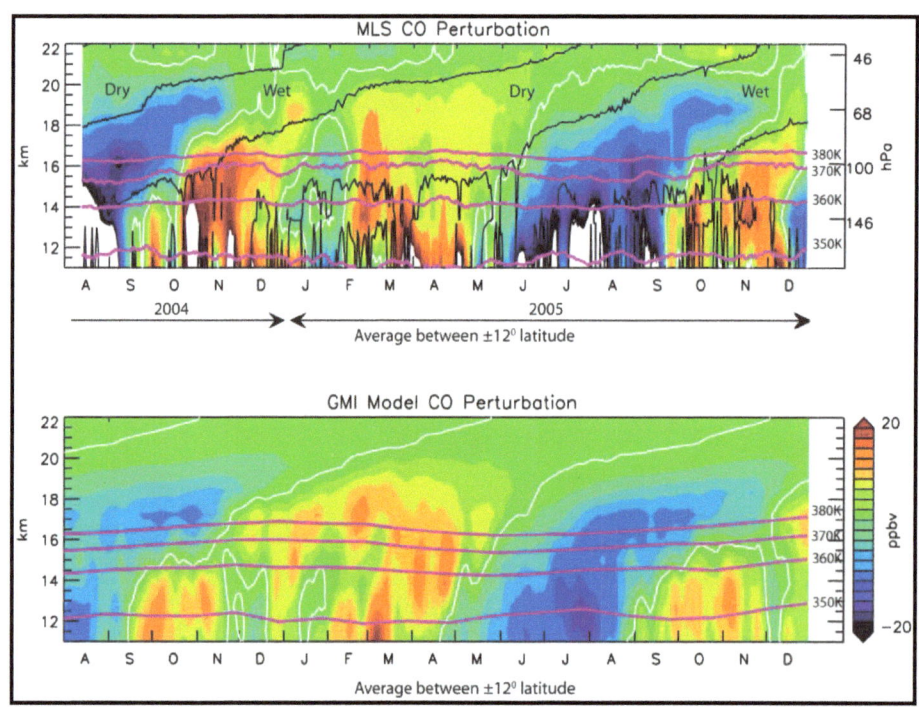

Dennis Flower manages MLS, which was designed and built at JPL.

122

The Tropospheric Emission Spectrometer (TES) on Aura

The Tropospheric Emission Spectrometer (TES) is designed to measure the composition of Earth's troposphere, the layer of the atmosphere that we breath, extending from Earth's surface to about 6 miles in altitude.

TES is a spectrometer that measures the infrared-light energy (radiance) emitted by Earth's surface and by gases and particles in Earth's atmosphere. Every substance emits infrared radiation at certain signature wavelengths. Spectrometers measure this radiation as a means of identifying the substances.

TES has very high resolution, which gives it the ability to precisely identify substances, and also provides information about their location in the atmosphere.

TES can detect and measure many components of the atmosphere, but one of its main purposes is to study ozone. High levels of ozone in the troposphere are usually associated with polluted environments, and are dangerous to both plants and animals, including humans. (The "ozone hole" is a concern in the upper atmosphere, or stratosphere.)

TES is providing important data on where the ozone in the troposphere comes from and how it interacts with other chemicals in our air.

The image to the right shows TES as it is being assembled.

TES is one of four instruments that fly on GSFC's Aura spacecraft. Aura's mission is to measure trace gasses in the atmosphere. This data allows scientists to better address global climate change issues such as global warming, the global movement, distribution and chemistry of polluted air, and ozone.

TES observes both straight down (nadir view) and at a sideways angle (limb view) behind the satellite. Limb viewing provides a much longer path through the atmosphere, and looking through a larger mass of air improves the chances of observing sparsely distributed substances that might be missed in the nadir view. Limb viewing's sideways angle also makes it easier to determine the altitudes of the observed substances. But limb viewing is

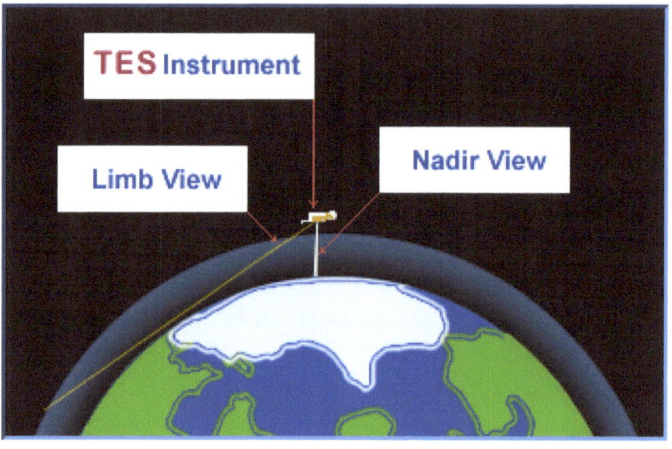

very susceptible to interference (only rarely does the line of sight reach the surface). Nadir viewing is less impacted by clouds, but looking straight down makes it more difficult to identify altitudes.

The image below shows measurements taken on September 20, 2004 of ozone concentrations in the lower troposphere. The map covers the Earth from 82° N to 82° S.

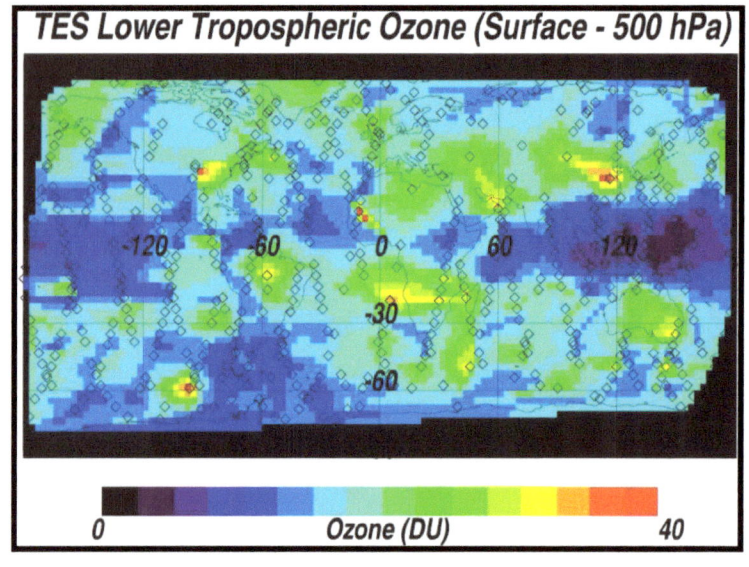

The color scale shows the number of ozone molecules per square centimeter of air in a column of atmosphere.

The numbers on the maps show latitude and longitude. The red, orange and yellow colors show high ozone levels.

TES feeds data about ozone concentrations to local agencies and decision-makers. Since TES is the only remote-sensing instrument currently flying that can distinguish ozone at altitudes where it does harm from altitudes where it is beneficial, TES data can be uniquely valuable to local and regional air-pollution analysis and forecasting.

Greenhouse gases contribute to global warming by trapping some of the energy that Earth radiates after being warmed by the sun. TES scientists are currently able to calculate how much energy is trapped at various altitudes by ozone and they're in the process of doing the same with water vapor, the most abundant greenhouse gas. This is basic research that will enable scientists to understand climate change.

The TES project was managed by Tom Glavich, and was designed and built at JPL.

Deep Impact – Striking a Comet

Launched in January 2005, NASA's Deep Impact spacecraft traveled about 268 million miles to the vicinity of comet Tempel 1.

On July 3, 2005, the spacecraft deployed an impactor that was essentially "run over" by the nucleus of comet Tempel 1 on July 4. This deep space collision took place at the speed of about 23,000 miles per hour.

Before, during and after the demise of this 820-pound impactor, a "flyby" spacecraft watched the 4-mile wide comet nucleus from nearby, collecting pictures and data of the event.

The impactor's impact with comet Tempel 1 formed a large crater, with ice and dust debris ejecting from the crater.

Competitively selected under NASA's Discovery Program, the project was managed by JPL and the spacecraft was built by Ball Aerospace. The picture to the left shows the spacecraft being assembled at a Ball facility.

Technical Challenge
The primary technical challenge was to navigate the impactor to achieve a very high speed interception. The relative speeds greatly exceed the speed involved in intercepting an ICBM, which was considered by many at that time to be impossible. This entire project was bid at less than $300 million, including the launch vehicle. There would be one shot and one shot only to pull off the intercept.

Programmatic Challenge

Although it was competitively selected in 2000 as a "faster, better, cheaper" type of mission, programmatic rules changed, due in part to the MCO and MPL failures. There was about a 30% cost overrun, roughly half of which was needed to overcome technical challenges, and the rest needed to address schedule delays and additional activities to respond to new programmatic requirements.

The Result

The mission was a great success. The image below shows comet Tempel 1, 67 seconds after it obliterated Deep Impact's impactor spacecraft. The image was taken by the high-resolution camera on the mission's flyby craft. Scattered light from the collision saturated the camera's detector, creating the bright splash seen here. Linear spokes of light radiate away from the impact site, while reflected sunlight illuminates most of the comet surface. The image reveals topographic features, including ridges, scalloped edges and possibly impact craters formed long ago.

The spacecraft that delivered the impactor and monitored the collision was then returned to the vicinity of Earth.

It is now being re-used for another mission, EPOXI

The project was managed by Brian Muirhead and then Rick Grammier. The spacecraft was built by Ball Aerospace Corporation.

Mars Reconnaissance Orbiter – Investigating Mars

Mars Reconnaissance Orbiter (MRO) is among the most capable planetary orbiters that JPL has developed so far. During its two-year primary science mission, the Mars Reconnaissance Orbiter has conducted eight different science investigations at Mars. The investigations are divided into three purposes: global mapping, regional surveying, and high-resolution targeting of specific spots on the surface.

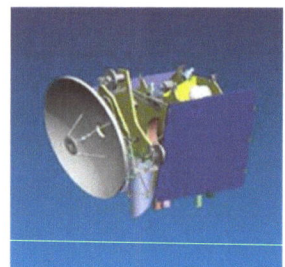

Because the spacecraft had to fit within the nose cone, or payload fairing, during launch, large parts like the high-gain antenna and the solar arrays were designed to be folded up. As soon as the launch vehicle put the spacecraft on course to leave Earth orbit for its journey to Mars, it disconnected itself from the spacecraft. The picture to the left shows this configuration.

As soon as the spacecraft was clear of the launch vehicle, the orbiter deployed its solar arrays to begin producing power. The high-gain antenna was also deployed at this point. The high-gain antenna moved to track the Earth, while the solar panels remained fixed. The picture to the right shows this configuration, as viewed from Earth.

During the primary science phase, the orbiter's job is to point its science instruments at Mars to collect images and other data from targets on the surface of Mars, while ensuring that the high-gain antenna and solar arrays are continuously tracking the Earth and the Sun, respectively. The picture to the left shows this configuration.

JPL's Jim Graf managed MRO. Lockheed Martin developed the spacecraft. Instruments were provided by Johns Hopkins University, Applied Physics Laboratory, Malin Space Science Systems, Univ. of Arizona and Ball Aerospace & Technologies Corporation, JPL, and the Italian Space Agency.

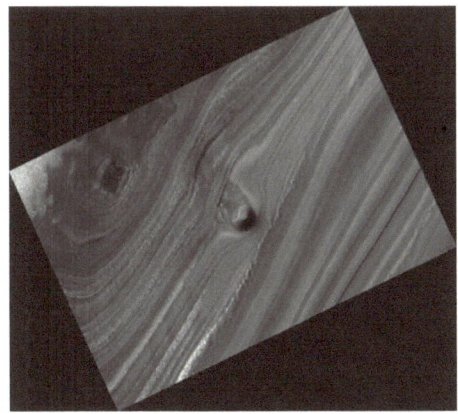

This striking image from the High Resolution Imaging Science Experiment camera shows a mound within the area of a trough cutting into Mars' north polar layered deposits. The camera took this image on Sept. 2, 2008.

The north polar layered deposits are a stack of layers that are rich in water-ice. The stack is up to several miles thick. Each layer is thought to contain information about the climate that existed when it was deposited. We can see these internal layers exposed where erosion has cut into the stack. One of these troughs is shown in this image and contains a 1,640-foot thick section of the layering.

A conical mound partway down the slope stands approximately 130 feet high. One possible explanation for this unusual mound is that it may be the remnant of a buried impact crater now being exhumed. **Image Credit**: NASA/JPL-Caltech/Univ. of Arizona

Images acquired by the orbiter reveal that different layers of rock have different properties and chemistry. The opal minerals are located in distinct beds of rock outside of the large Valles Marineris canyon system and are also found in rocks within the canyon. The presence of opal in these relatively young rocks tells scientists that water, possibly as

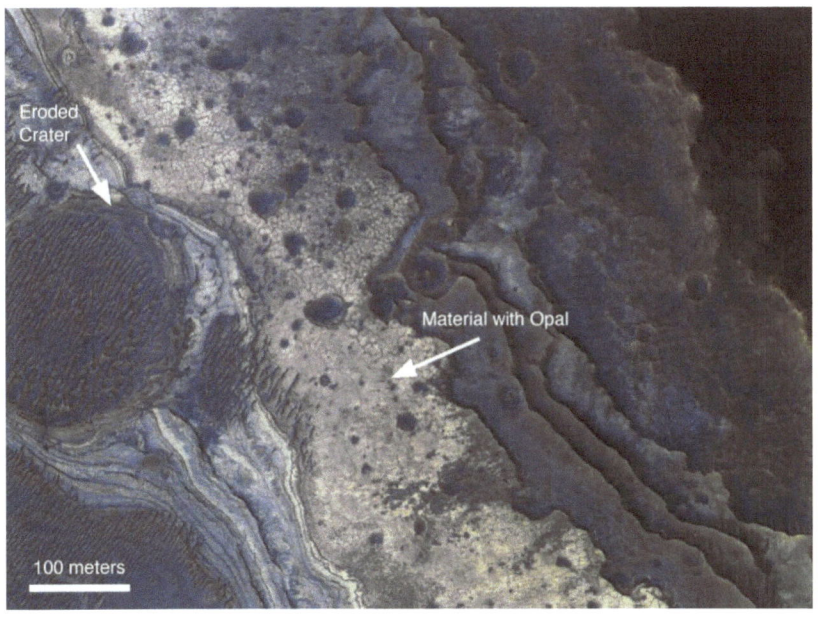

rivers and small ponds, interacted with the surface as recently as two billion years ago, one billion years later than scientists had expected. The discovery of this new category of minerals spread across large regions of Mars suggests that liquid water played an important role in shaping the planet's surface and possibly hosting life. Image Credit: NASA/JPL-Caltech/Univ. of Arizona

Images (a) through (d) below show the landing site of the Mars Exploration Rover Spirit.

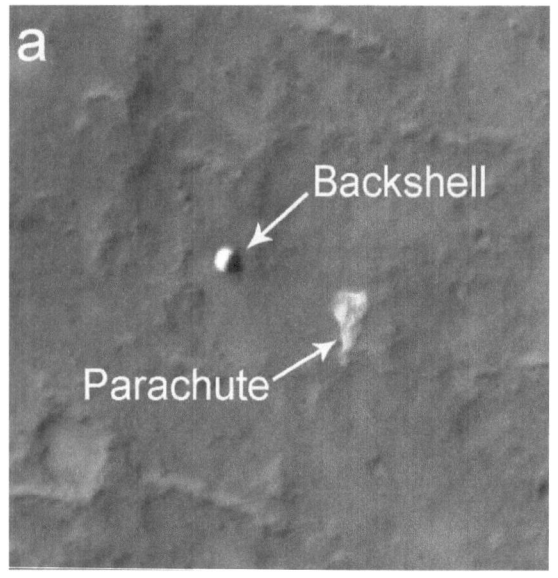

The bright irregularly shaped feature in area "a" of the image is Spirit's parachute, now lying on the Martian surface.

Near the parachute is the cone-shaped back shell, which helped protect Spirit's lander during its seven-month journey to Mars. The back shell appears relatively undamaged by its impact with the Martian surface. Wrinkles and folds in the parachute fabric are clearly visible.

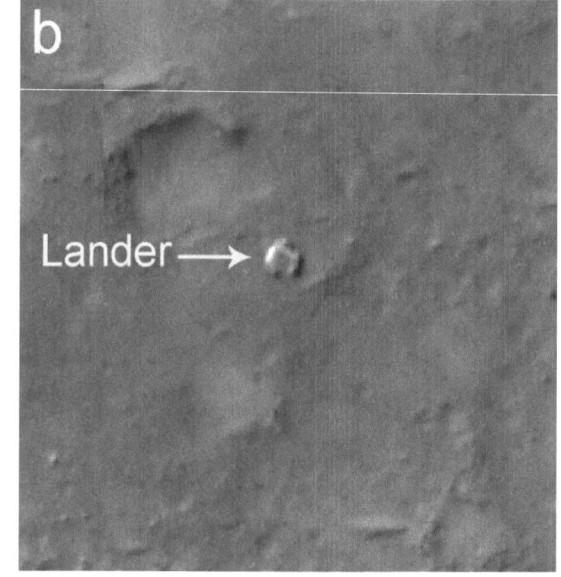

Area "b" of the image shows Spirit's lander.

The crater in the upper left portion of the image, just northwest of the lander, was informally named "Sleepy Hollow" by the Mars Exploration Rover team.

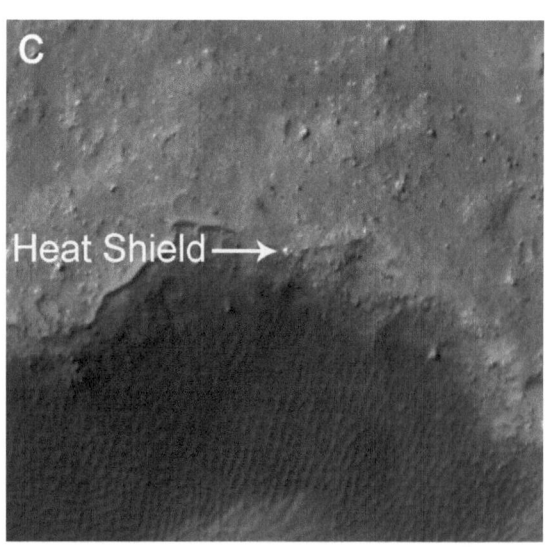

Area "c" of the image shows Spirit's heat shield at the edge of Bonneville Crater.

Area "d" of the image shows the location of Spirit on September 29, 2006. Toward the top of the image is "Home Plate," a plateau of layered rocks that Spirit explored during the early part of its third year on Mars.

Spirit itself is clearly seen just southeast of Home Plate. Also visible are the tracks made by the rover before it arrived at its current location.

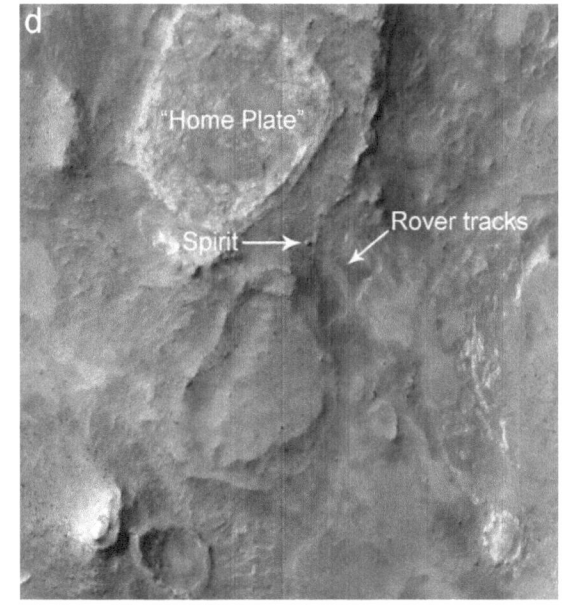

MRO completed its primary mission in November, 2008. It is expected to continue to operate and to return scientific information about Mars until at least December, 2010.

The Microwave Instrument (MIRO) on the Rosetta Orbiter

This JPL instrument is a combination spectrometer and radiometer. MIRO will study gases given off by comet 67P/Churyumov-Gerasimenko as part of the payload on the European Space Agency's Rosetta spacecraft.

MIRO will measure four of the ten most abundant types of molecules in a comet nucleus (water, carbon monoxide, ammonia, and methanol). It will also help determine how the comet's surface and subsurface temperatures change as it approaches the Sun.

The Rosetta/MIRO timeline:

- In March 2004, Rosetta took off on its ten-year voyage.
- In March 2005, Rosetta rocketed around Earth, using our planet as a gravity assist to gain the momentum it needed to reach Mars.
- On July 4, 2005, Rosetta observed JPL's Deep Impact spacecraft as it sent an "impactor" into the path of Comet 9P/Tempel-1. From 46 million miles away, Rosetta's instruments, including MIRO, detected the chemical composition of the gas jetting from the explosion. The results surprised scientists: Tempel-1 is not the icy rock they expected. It's more like a "souffle" of dust containing only about six percent pure water ice.
- Rosetta flew around Mars on February 25, 2007, getting as close as 100 miles.
- On November 13, 2007, Rosetta swung by Earth for the second time, using gravity assist to reach Asteroid Steins in September, 2008.
- In November 2009, Rosetta returns to Earth for its third and final flyby. Rosetta will use this final boost from Earth's gravity to reach asteroid Lutetia (2010), and then attain its farthest point from the sun, where it will hibernate for almost three years before resuming its mission.
- In January 2014, Rosetta will awake and begin its drop back down to the inner solar system for its rendezvous with the comet in May 2014. Over the following three months, Rosetta's rocket thrusters will slow it to a speed of just two meters/second relative to the comet. Rosetta will then insert itself into the comet's orbit, deploy a lander, and allow MIRO and other instruments to study comet 67P/Churyumov-Gerasimenko as it approaches the sun.

So far, every ambitious step of this mission has worked out as planned.

CloudSat – Profiling Earth's Clouds

CloudSat is the first satellite that uses an advanced radar to "slice" through clouds to see their vertical structure, providing a completely new observational capability from space (previous weather satellites could only image the uppermost layers of clouds). CloudSat furnishes data needed to evaluate and improve the way clouds are represented in global models, thereby contributing to better predictions of clouds and thus to their poorly understood role in climate change and the cloud-climate feedback.

CloudSat is an interagency mission with project management by Tom Livermore (through launch) and Deborah Vane (through operations) of JPL. Partners include the University of Colorado, the Canadian Space Agency, the U.S. Air Force and the U.S. Department of Energy.

Ball Aerospace built the spacecraft, which can be seen to the left. JPL built the radar.

CloudSat flies in tight formation with the CALIPSO satellite, and these two satellites follow behind the Aqua satellite in a somewhat looser formation.

This is informally called the "A-Train", where "A" stands for Aqua. Other satellites have joined the "A-Train", as can be seen in the diagram below.

The combination of data from the CloudSat radar with coincident measurements from CALIPSO, Aqua and the other A-Train satellites provide a coordinated set of information that can be used to study weather and climate.

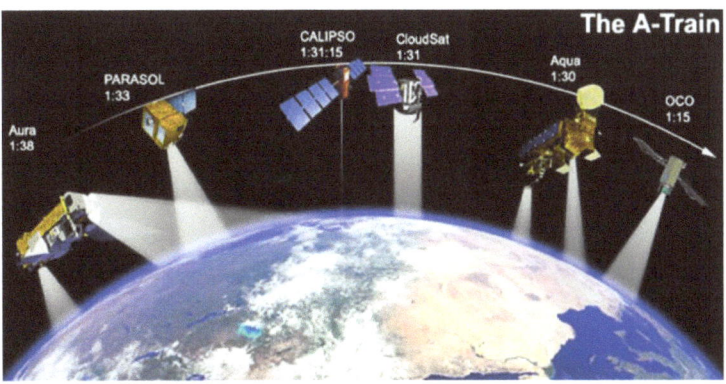

A separate CloudSat study led by John Haynes at Colorado State University found that it rains more often and in greater amounts over Earth's oceans than previously estimated. On average,

13 percent of clouds over Earth's oceans produce rain that reaches the surface. The difference between prior estimates and actual measurements was greatest during winter.

"These results suggest there is considerably more water falling from our skies, at least over Earth's oceans, than we previously thought," said Haynes. "The implications of these results are substantial and are still being examined, and suggest it may be necessary to reassess climate model estimates of Earth's water cycle intensity. By improving our understanding of present rainfall patterns, scientists can also improve climate model projections of how rainfall will increase or decrease in the future around the world."

CloudSat is providing some of the first, most direct observations of where rainfall occurs on a near-global basis, allowing scientists to see, for the first time, what fraction of Earth's clouds precipitate. It surveys regions where measurements did not previously exist.

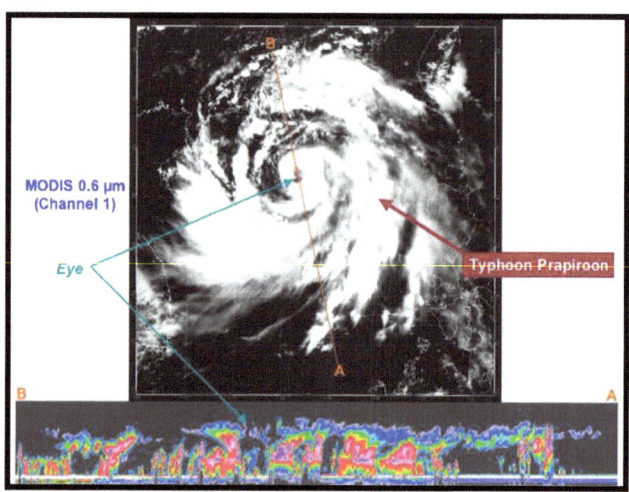

At approximately 0553 UTC (1:53 am EDT), on 2 Aug 2006, CloudSat flew over the eye of Typhoon Prapiroon as it approached southern China. The upper image is from the MODIS instrument on the NASA Aqua satellite, to give an idea of how the storm looked from the top. The bottom image is from NASA's new CloudSat satellite. The CloudSat radar flies in an on-orbit formation with the Aqua satellite, approximately one minute behind, as part of the A-Train constellation of satellites. The red and purple areas indicate large amounts of cloud water. The blue areas along the top of the clouds indicates cloud ice, while the wavy blue lines on the bottom of the image indicate intense rainfall.

Here is a profile of a thunderstorm.

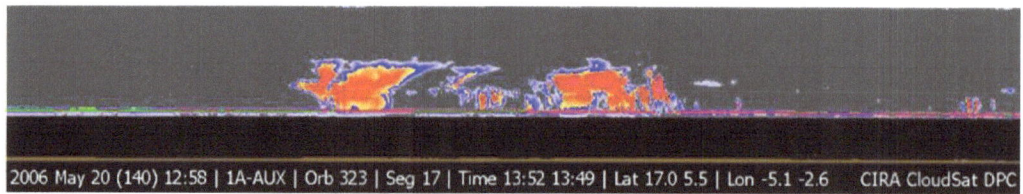

Results
CloudSat performed like a champion, returning over 99.9% of the possible science data, at levels of precision and accuracy that are 1,000 times greater than earlier technologies.

EPOXI and NEXT – Reusing Deep Space Spacecraft

Both EPOXI and NEXT are extended missions, using the Deep Impact and the Stardust spacecraft, respectively. One management team runs both missions, led originally by Tom Duxbury, and currently by Tim Larson.

EPOXI

Sixteen days after comet encounter, the Deep Impact team placed the spacecraft on a trajectory to fly past Earth in late December 2007. The spacecraft was then reconfigured for a new mission called EPOXI.

EPOXI utilizes the already "in flight" Deep Impact spacecraft to explore distinct celestial targets of opportunity. The name EPOXI itself is a combination of the names for the two extended mission components: the extrasolar planet observations, called Extrasolar Planet Observations and Characterization (EPOCh), and the flyby of comet Hartley 2, called the Deep Impact Extended Investigation (DIXI). Note - while the mission name has been changed to EPOXI the spacecraft will continue to be referred to as "Deep Impact."

In early 2008 the EPOXI science team used the spacecraft's large telescope to observe five nearby stars with "transiting extrasolar planets," so named because the planet transits, or passes in front of, its star. The Earth and the Moon were used for practice for this stage of the mission. Here is a series of four images showing the Moon transiting Earth, captured by EPOXI.

 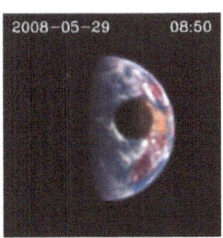 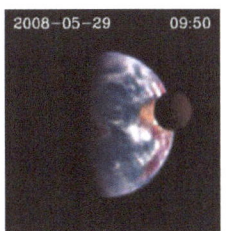 

Image Credit: Donald J. Lindler, Sigma Space Corporation/GSFC; EPOCh/DIXI Science Teams.

Then the spacecraft trajectory was altered by a December 29, 2008 Earth flyby, to put it on course to intercept the comet Harley 2. The spacecraft will be put into "hibernation" only wake up in the fall of 2010 when it arrives in the vicinity of Hartley 2. On November 4, 2010 EPOXI will conduct an extended flyby of Hartley 2 using all three of the spacecraft's instruments (two telescopes with digital color cameras and an infrared spectrometer).

NEXT

NEXT will use the Stardust spacecraft to conduct a very close (200 mile) flyby of the nucleus of comet Tempel 1. It is expected to get image resolution of 12 meters per pixel, and to get accurate measurements of the size, composition, and distribution of dust particles surrounding the nucleus of this comet.

Tempel 1 is the same comet that was previously impacted by Deep Impact. This is an opportunity to get a better look at what happened, and to explore material ejected from within the comet.

This mission is made possible by an Earth gravity assist maneuver in January, 2009. The Tempel 1 flyby will occur in February, 2011. An image from the Solar System Simulator shows where it is at the time this was written.

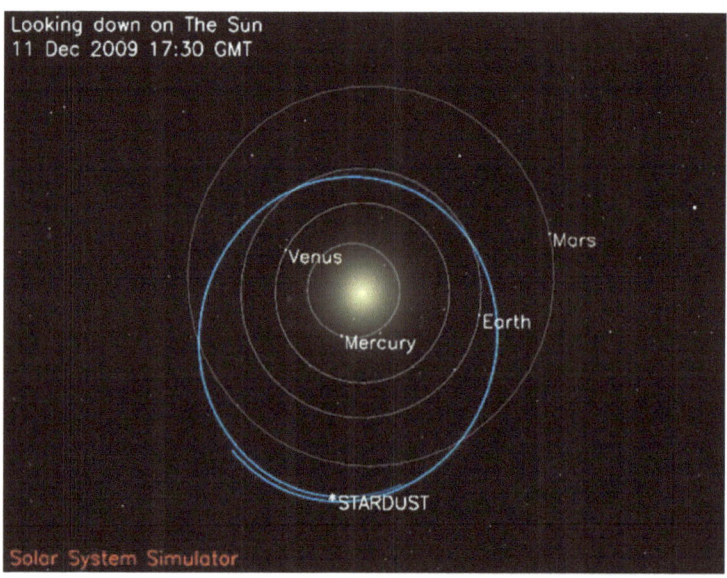

Phoenix – Putting a Laboratory on Mars

The original Mars 2001 launch was going to have both an orbiter (Odyssey) and a lander. Due to the failure of MPL and MCO, the lander part of the mission was postponed.

The Phoenix mission finally delivered that lander to a region close to the North Pole of Mars. Remote instruments had indicated that there was ice, right under the surface of Mars in this region. Phoenix would (1) land on Mars, (2) a robot arm would dig down under the surface as much as 2 feet, and (3) move samples of whatever was down there into a portable laboratory for analysis.

The scientific purpose was to: (1) confirm the presence of water, if possible, and to identify other chemicals under the surface; (2) determine if organic molecules existed in any of the samples tested; and (3) monitor weather conditions on Mars.

The mission was selected in August, 2003. The University of Arizona provided the Principal Investigator, JPL's Barry Goldstein managed the mission and JPL built some of the instruments, Lockheed Martin built the spacecraft and the lander, and the Canadian Space Agency provided the meteorological instruments.

The Phoenix flight system is shown to the right. It is being prepared for testing in a vacuum chamber.

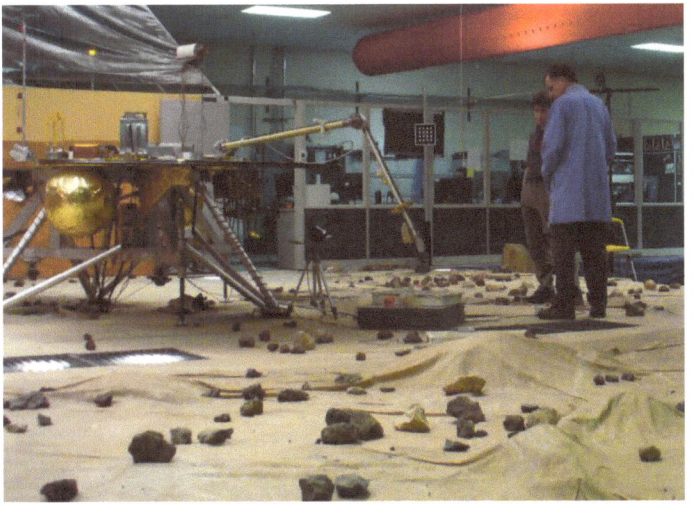

The lander is shown to the left, in a test area.

The lander hit the top of the Martian atmosphere at 12,750 miles per hour. The heat shield protected the lander as it slowed down, due to atmospheric drag. Then a parachute, and finally a set of thrusters were used to slow the lander down to a speed of a little over 5 miles per hour when it touched the surface.

Here is an early image of the surface. This image shows NASA's Phoenix Mars Lander's solar panel and the lander's Robotic Arm with a sample in the scoop.
Image credit: NASA/JPL-Caltech/University of Arizona/Texas A&M University

The white areas seen in these trenches are part of an ice layer beneath the soil. The trench on the upper left is about 24 inches long and 1 inch deep. The trench in the middle is about 12 inches long and 1 inch deep. The trench on the right is about 12 inches long and 2 inches deep.

Here is a picture of the Martian terrain around the lander.

Phoenix was designed to last for 50 Martian days (Sols). A Martian day is 40 minutes longer than an Earth day. This is among the last images of the Martian terrain taken by Phoenix on Sol 151 of the mission (October 27, 2008).

Dawn – Exploring Our Solar System's Past

Dawn investigates the internal structure, density and homogeneity of two complementary protoplanets [Protoplanets in this context are moon-sized planets], 1 Ceres and 4 Vesta.

It is believed that only the largest asteroids remain relatively undisrupted by the passage of comets and the influences of planets.

The most massive of these are Ceres and Vesta. The former has a primitive surface, with water-bearing minerals, and possibly a very weak atmosphere and frost. The latter is a dry, differentiated body whose surface has been resurfaced by basaltic lava flows possibly possessing an early magma ocean like the Moon.

To the left is an image of Ceres. Credit: Keck Observatory by C. Dumas. To the right is an image of Vesta. Credit: Ben Zellner (Georgia Southern Univ.) and Peter Thomas (Cornell University & NASA)

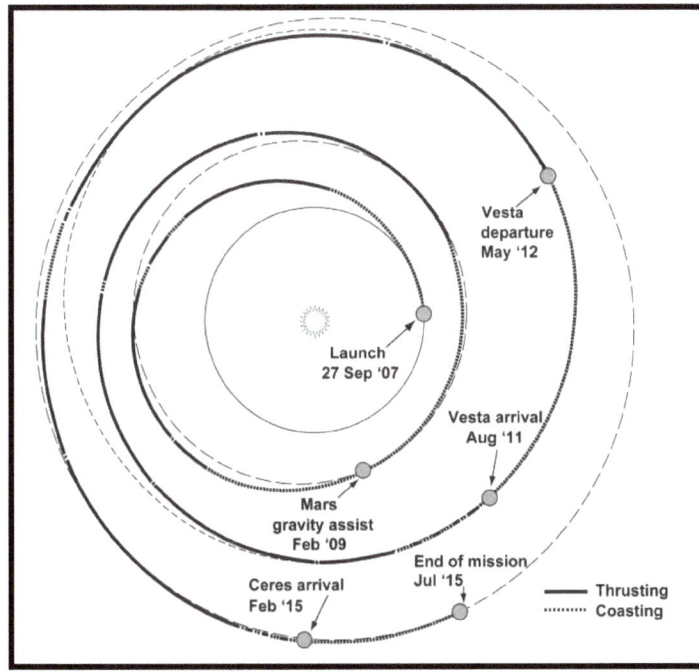

A unique technical feature of this mission is the challenge of going into orbit around these relatively small bodies. The planned trajectory is shown to the left.

As you can see, a Mars gravity assist occurs in February, 2009.

Dawn will orbit both Vesta and then Ceres for several months.

Dawn experienced significant management turnover. The first project manager was Sarah Gavit. She was followed by Tom Fraschetti and finally by Keyur Patel, who took the project to the launch pad.

A Funny Thing Happened to Our Solar System in 2006

The term "protoplanet" may be new to some readers. In this context, a "protoplanet" is a moon-sized planet.

You might be interested to learn that the International Astronomical Union has determined that the number of planets in our Solar System is no longer 9, but either 8 or 11 or more. Over a period of some years, this body of experts has classified Pluto as a protoplanet, then changed its mind and created a definition of "planet" that includes such large objects as Ceres, (previously referred to as an asteroid).

This is due in part to recent research by Professor Michael Brown of Caltech, who discovered UB313 in 2005. Prof. Brown described UB313 as the "tenth planet". UB313 is larger than Pluto, but much farther out, in what is known as the Kuiper belt of the Solar System. Image credit: Caltech.

It is almost 10 billion miles from the sun and more than 3 times more distant than Pluto (sometimes) and takes more than twice as long to orbit the sun.

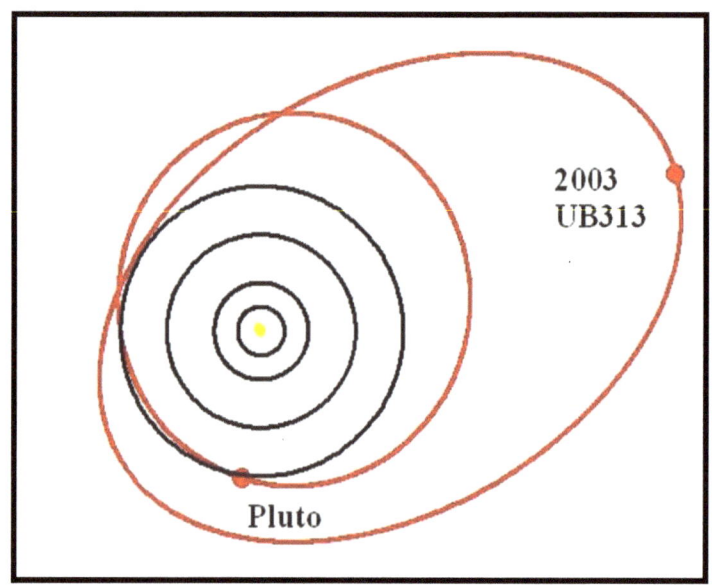

The International Astronomer's Union held a meeting to determine if this new discovery should be called a planet, or something else. At first they decided to call UB313 and all similar objects "protoplanets", and they moved Pluto in this new category. In order to be consistent, anything inside the solar system that was roughly a moon-sized object that was not orbiting an actual planet, was also called a "protoplanet". So Ceres became a protoplanet, instead of just an asteroid.

Unfortunately, many people did not like the demotion of Pluto from "planet" to "protoplanet". The International Astronomical Union went back to work and came up with a new definition of "planet" that would include Pluto, Ceres, and UB313. Then they also came up with an official name for UB313, Eris. In Greek mythology, Eris is the goddess of warfare and strife. This part of the story makes sense, when you consider all the confusion that emerged when Eris was discovered.

Jason 2 – Continuing the Legacy of TOPEX

This mission was originally called the Ocean Surface Topography Mission, and is now called Jason-2. It is a partnership between NASA, the National Oceanic and Atmospheric Administration (NOAA), the French Space Agency and the European Organization for the Exploitation of Meteorological Satellites. JPL's work was managed by Parag Vaze. Jason 2 will extend into the next decade the continuous record of sea-surface height measurements started in 1992 by the NASA-French Space Agency's TOPEX/Poseidon mission and extended by the NASA-French Space Agency's Jason 1 mission in 2001.

After all the instruments have been successfully integrated and tested, the spacecraft engineers conduct performance verification tests on the entire payload to make sure that the instruments work together as specified.

At different times during system level tests, the entire spacecraft, including its payload, is shocked, shaken, frozen and cooked, all in an effort to ensure that the spacecraft will survive the violence of the launch and will operate in the harsh environment of outer space.

This image shows the finished product of the assembly process.

The Jason 2 launched on June 20, 2008. It is planned to operate for at least five years.

Pools of warm water known as Kelvin waves can be seen traveling eastward along the equator (black line) in this image from the Ocean Surface Topography Mission/Jason-2 satellite. El Niños form when trade winds in the equatorial western Pacific relax over a period of months, sending Kelvin waves eastward across the Pacific like a conveyor belt.

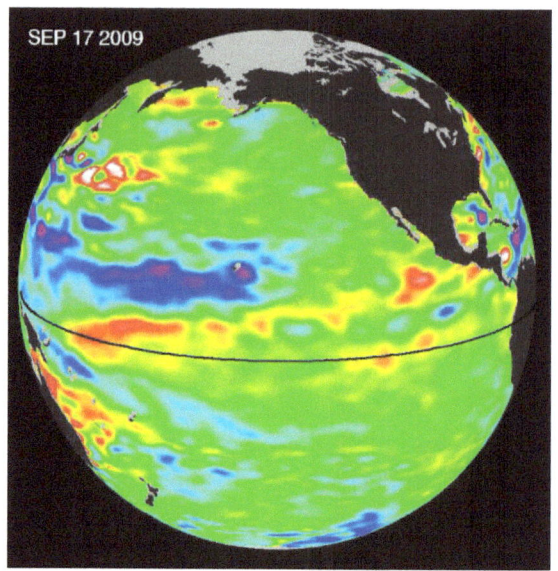

Since May 2009, the tropical Pacific Ocean has switched from a cool pattern of ocean circulation known as La Niña to her warmer sibling, El Niño. This cyclical warming of the ocean waters in the central and eastern tropical Pacific generally occurs every three to seven years and is linked with changes in the strength of the trade winds.

Long-term weather patterns influence water supply, food supply, trade shipments, and property values. Over the past 50 years, the ocean has absorbed over 80 % of the heat from global warming. Ocean surface topography data have revolutionized our understanding of the ocean. Some of the important scientific results include:

- Confirmation and accurate measurements of global sea level rise
- Global- and decadal-scale observations and analyses of El Niño and La Niña
- New discoveries in ocean circulation and its effects on climate
- Improvements in ocean tide models

Jason-2 will extend the length of the sea surface height measurement record to two decades, improving our understanding of long-timescale climate events such as the Pacific Decadal Oscillation. This two-phased pattern of climate variability can last from 10 to 70 years, much longer than El Niño or La Niña events.

The Moon Mineralogy Mapper on Chandrayaan-1

The Moon Mineralogy Mapper (M^3) is a JPL instrument that is part of India's first mission to the Moon, Chandrayaan-1 (meaning "Lunar Craft" in ancient Sanskrit). The Indian Space Research Organization launched Chandrayaan-1 on October 22, 2008, from Sriharikota, India. It entered lunar orbit on November 8.

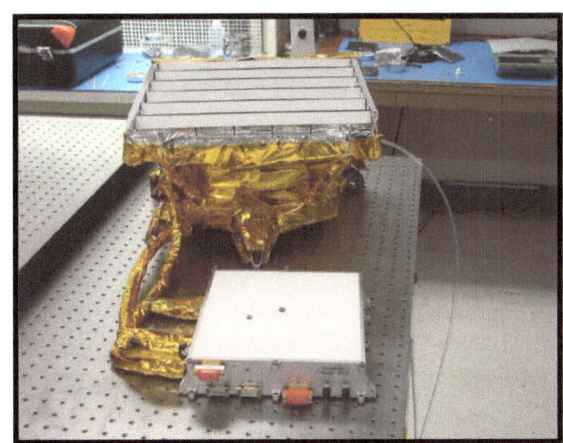

M^3 is a state-of-the-art imaging spectrometer that is now providing the first high resolution map of the entire lunar surface, revealing the minerals of which it is made.

Scientists will use this information to answer questions about the Moon's origin and development and the evolution of terrestrial planets in the early solar system. Future astronauts will use it to locate resources, possibly including water, which can support exploration of the Moon and beyond.

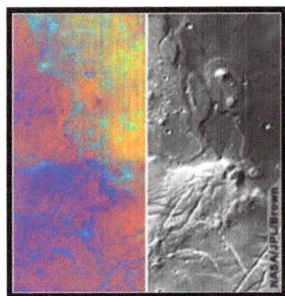

The composite image to the left shows M^3 data for the Orientale region of the moon. The strip on the left is a composite of data from 28 separate wavelengths of light reflected from the moon. The blue to red tones reveal changes in rock and mineral composition, and the green color indicates the abundance of iron-bearing minerals. The image on the right is from a single wavelength for thermal emission, providing a new level of detail on the form and structure of the region's surface.

Image Credit: NASA/JPL/Brown

M^3 was managed by Tom Glavich and then by Mary White. M^3 was cited by NASA as one of its top 10 accomplishments for 2008. The M^3 team plans to continue their mission of lunar exploration for at least the next two years.

JPL Flight Projects that will Launch after 2008

This book describes fifty years of mostly successful flight projects at JPL, which launched from 1958 to 2008. Here is a brief mention of flight projects that are being designed or built right now, in 2009. Updates are regularly posted at the JPL public website, at www.jpl.nasa.gov.

Orbital Carbon Observatory (OCO)
- Launched in February, 2009
- OCO was designed to measure the CO_2 in Earth's atmosphere, using a high resolution spectrometer (see picture to the right)
- Sadly, as this book was nearing completion in 2009, the OCO mission was lost, apparently due to a failure in the launch system. A tiger team has been assembled to propose a replacement mission for this important scientific experiment.

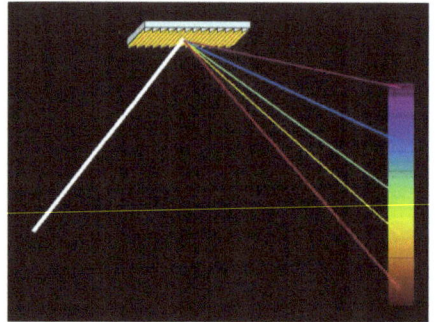

Kepler
- Launched in March, 2009
- Kepler will survey distant stars to find Earth-sized planets in our Galaxy
- Kepler has a telescope that measures the brightness of 100,000 stars, and will do so for 4 years without interuption. When a planet crosses in front of its star, the change of brightness gives the size of the planet and time between transits provides the orbit
- The challenge now is to find terrestrial planets in the habitable zone of their stars where liquid water and possibly life might exist
- It has already detected the atmosphere of a known giant gas planet, demonstrating the telescope's extraordinary scientific capabilities.

Diviner instrument, on the Lunar Reconnaissance Orbiter

- Launched on Goddard's Lunar Reconnaissance Orbiter in June, 2009
- Diviner will measure lunar surface temperature profiles, characterize thermal environments for habitability, and perform other surface investigations.

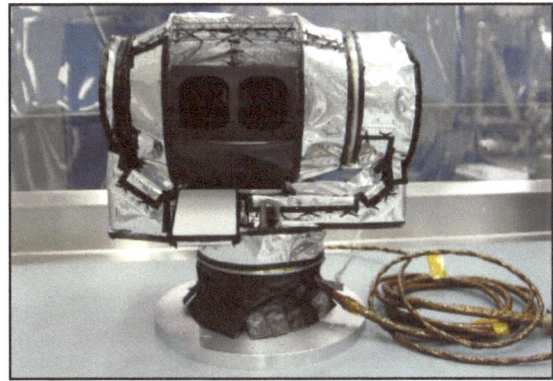

Herschel instrument on the European Herschel Space Observatory

- Launches in 2009
- Herschel provides components for the Observatory instruments

Planck instrument on the European Planck mission

- Launches in 2009 along with Herschel
- NASA' Planck project provides components for this mission's instruments

Widefield Infrared Survey Explorer (WISE)

- Launches in late 2009 or early 2010
- WISE will be an infrared telescope that has 500 times greater sensitivity than IRAS.
- It will:
  - Characterize albedos and sizes of hazardous asteroids
  - Examine the nearest planetary system (a brown dwarf)
  - Identify the most luminous galaxies in the Universe.

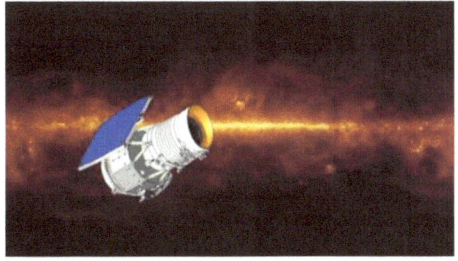

Aquarius

- Launches in 2010
- Aquarius measures Sea Surface Salinity
- This project is a collaboration between Argentina, JPL and GSFC.

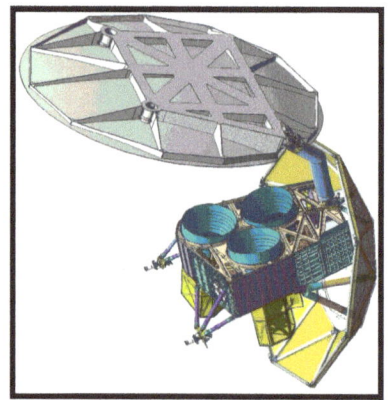

Juno

- Launches in August, 2011
- This mission will conduct an in-depth study of the giant gas planet Jupiter
- Juno is the first solar powered mission to Jupiter
- It will conduct gravity, magnetic and atmospheric investigations.

NuSTAR

- Launches in 2011 (see image below)
- NuSTAR will observe hard X-rays to detect black holes of all sizes and other exotic phenomena
- NuSTAR has more than 500 times the sensitivity of previous instruments that detect black holes.

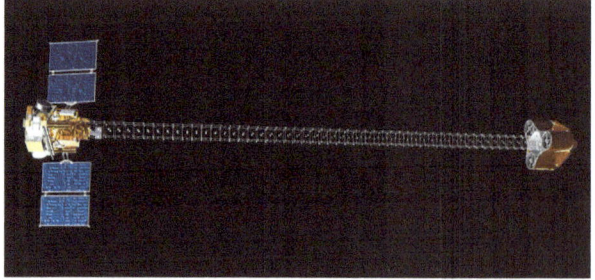

GRAIL
- Launches in 2011
- GRAIL will fly twin spacecraft in tandem orbits around the moon to measure its gravity field. The payload is derived from the GRACE mission, and is designed to determine the moon's interior structure, and to investigate its structural and thermal evolution

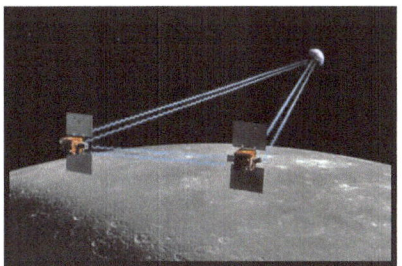

Mars Science Laboratory (MSL)
- Launches in October, 2011
- MSL is the next generation of Mars rover. It will
  - Collect soil and rock samples and analyze them for organic compounds
  - Study geologic processes that formed these rocks
  - Study the Martian atmosphere
  - Determine the distribution of water and carbon dioxide.

Soil Moisture, Active and Passive (SMAP)
- Is just starting up
- Launches in 2015
- SMAP will provide soil moisture maps over much of the land surface of Earth every 2 or 3 days
- The author of this book is currently working on SMAP
- Artist's concept is shown at right.

Mid Infrared Instrument (MIRI) on the James Webb Space Telescope
- MIRI will provide a 100 times advance in sensitivity over previous telescope/instrument combinations

Information Sources

The JPL public website at jpl.nasa.gov provides pages for all missions and many of the instruments mentioned in this book.

Other sources:
- "*Pathway to the future, the Jet Propulsion laboratory's implementation Plan*", May, 2003
- The JPL Fact Sheet
- The JPL Beacon Photo Index
- JPL Public Web Site, "Photojournal"
- Clayton R. Koppes, *JPL and the American Space Program*, Yale University Press, 1982
- The JPL Project Reference List, as of January, 2009 (a JPL-proprietary database on JPL projects compiled by the Author)
- NASA Facts for the following missions: Rangers and Surveyors to the Moon; Mariner to Mercury, Venus and Mars; Viking Mission to Mars; Magellan; Mars Global Surveyor; Mars Pathfinder; Spitzer; Epoxi and NExT; Phoenix; Dawn; and Jason 2
- Peter Westwick, *Into the Black, JPL and the American Space Program 1976-2004*, Yale University Press, 2007
- NASA, "*Final Report of the Ranger Board of Inquiry,*" November 30, 1962
- Notes from Barbara Amago on a lecture, "Voyager – the Journey of a Lifetime" by Dr. Ed Stone, July 26, 2002
- Lee-Leung Fu and Benjamin bolt, "*Seasat Views Oceans and Sea Ice with Synthetic Aperture Radar*", JPL Publication 81-120, February 15, 1982
- NASA Press Releases
- Genesis Mishap Investigation Report, Volume 1, November 2005
- Prof. Michael Brown, Caltech recorded lecture of February 22, 2006
- Personal notes from conversations with numerous individuals, including: Tony Spear (manager, Pathfinder); Dave Lehman (manager, Deep Space One and NuSTAR); Sara Gavit (manager, Dawn); Ken Atkins (manager, Stardust); Jim Graf (manager, Quikscat); Chet Sasaki (manager, Genesis); Ab Davis (manager, GRACE); and numerous other engineers and managers at JPL.

About the Author

Robert Aster has two degrees from Stanford University, and has worked at Caltech's Jet Propulsion Laboratory since 1976.

Robert has enjoyed several careers with one employer. His first career at JPL was in renewable energy research from 1977 to 1983. His second career at JPL was analysis of the performance of information systems, in the 1980s (See the chapter: "A Modest Contribution".). Robert's third career at JPL started in 1990 when he took up flight project planning.

Since 1990, Robert has been a leader in developing the JPL planning system, and has trained over a thousand engineers on how to plan and manage their work. He also led the initial development of JPL's system for strategically planning its technical workforce. In this era, JPL grew from managing several flight projects at a time to managing nearly thirty flight projects.

In recent years, Robert also contributes material to JPL's new mission proposals. Some of that material can be found in this book.